YOUR GUIDED TOUR TO THE
FEDERAL BUDGET CRISIS

SCOTT BITTLE
& JEAN JOHNSON

 Collins

An Imprint of HarperCollinsPublishers

HarperCollins books may be purchased for educational, business, or sales promotional use. For information, please write: Special Markets Department, HarperCollins Publishers, 10 East 53rd Street, New York, NY 10022.

FIRST EDITION

Designed by Nancy Singer Olaguera

Library of Congress Cataloging-in-Publication Data

Bittle, Scott.
    Where does the money go? : your guided tour to the federal budget crisis / Scott Bittle and Jean Johnson.—1st ed.
        p. cm.
    ISBN 978-0-06-124187-1
1. Budget deficits—United States. 2. United States—Appropriations and expenditures. 3. Public welfare—United States—Finance. 4. Budget—United States. I. Johnson, Jean. II. Title.
    HJ2052.B58 2008
    336.73—dc22

                                                      2007026403

08 09 10 11 12   ISPN/RRD   10 9 8 7 6 5 4 3 2

For our dads,
Mulford W. Bittle Jr. (1934–2000) and John J. Johnson (1917–2005)–
men who would never have left debts for their children to pay

CONTENTS

PREFACE

Where *Does the Money Go? Your Guided Tour to the Federal Budget Crisis* is exactly what it says it is—a straightforward explanation of what politicians, economists, think tanks, and lobbyists are arguing about when they fight about the federal budget. On the face of it, this might seem like a policy wonk, C-SPAN kind of problem—one that's better left to people who really enjoy this kind of thing. Unfortunately, those are exactly the people who have been handling the problem until now, and frankly, they haven't done a really great job.

The United States is seemingly addicted to spending more than it takes in. We've already piled up an unbelievable national debt. Even worse, we face truly gigantic expenses as the baby boomers begin to retire and need more health care. Today's problems will seem like a fender bender compared to the economic train wreck the country will face if we don't get the nation's finances under control. In fact, you may want to get a head start on feeling nostalgic for the 1990s and early 2000s. "These *are* the good old days," says Douglas Holtz-Eakin, a former head of the Congressional Budget Office who is widely admired for his expertise and nonpartisanship.

We've written this book because we believe that there are millions of Americans who are uneasy about where the country is headed and want to understand what the budget situation really is and what the options are. We are also convinced that there are millions of Americans who are tired of political leaders—Republicans and Democrats, liberals and conservatives—who bob and weave around this issue. We think there are plenty of citizens across the country who want to begin to figure this thing out for themselves. As authors, our job is to explain the problem and the options as clearly and fairly as we can.

A WORD ABOUT WHERE WE'RE COMING FROM

Where Does the Money Go? is not a book for policy makers and economists. It's a guide for people who care about where the country is going, but don't have the time or inclination to become budget experts. And for readers who do get a bit hooked on the topic—sometimes people do—we've provided lists of publications, organizations, and Web sites specializing in things budgetary in the appendix, so you can go forth and multiply. The country definitely needs more of you. For most of us, however, the first order of business is to grasp the essentials so your votes and your contributions can go to candidates who truly represent your interests.

Since we're presuming to explain this topic, and since it is complicated and controversial, we believe readers have a right to know where we're coming from. We both work for a nonprofit, nonpartisan research organization called Public Agenda where we write about public policy issues and conduct public opinion research (you can check out the organization at www.publicagenda.org). Consequently, both of us have spent a lot of time listening to very knowledgeable people talk about the federal budget, the deficit, the debt, and the aging of the baby boomers. We also have a lot of experience translating expert information into terms and concepts nonexperts can understand. It's been part of our jobs. Here's the approach we've taken in *Where Does the Money Go?*

★ First, we consider ourselves translators, not budget experts. We have done our homework on the issue (our reading list is in the appendix), and we've asked for and gotten advice from some of the most informed and intellectually honest people in the country. We've pulled that "best advice" together in one volume. We hope we've done it in an understandable and readable way.

★ Second, we're explainers, not advocates. You want to find "the man with the plan"? There are books, articles, and op-eds in the thousands pushing for specific solutions to the country's financial problems. That's not our purpose here. Instead, we've tried to lay out the information and the choices so you can come to your own conclusions about what should be done. Our last chapters contain tools enabling you to do just that—decide the matter for yourself. Naturally, there are areas where the facts and figures are in dispute and where experts disagree (it's just part of their job description to disagree). When this happens, we try to help you understand the nature of the disagreement. When there seems to be a pretty strong consensus of expert opinion one way or the other, we tell you that, too.

★ Third, we're optimists, not pessimists. As you will read in the following pages, the federal government has been spending more money than it takes in for some time now, and we have some huge financial commitments coming due. A lot of experts believe the country is taking grave risks with the economy and our future standard of living by not addressing the problem. Some of the predictions for what could happen ten or twenty years down the road are frightening. But there are steps the country can take to reduce these risks, and we believe there are solutions and compromises that would be acceptable to millions of Americans. We're also convinced that most people, once they understand what's at stake, will want to act and elect leaders committed to tackling this problem. As we said, we're optimists.

★ Fourth, we believe the budget problems are serious. You can certainly find experts who say otherwise, who believe we shouldn't worry about continuing deficits as long as the economy is growing, that the country's debt isn't that large given the size of the U.S. economy, or that the debt doesn't matter given the way global markets work today. You can also find experts who argue that we can grow our way out of these problems if we just make the right economic decisions. We've considered their arguments, and we discuss the key ones later. In many ways, it would be so nice to sit back, relax, and hope their predictions turn out just the way they say. But based on everything we've learned, we think it's just far too risky for the country to continue on its current merry financial path. We're optimists, but we're not prepared to cross our fingers, close our eyes, and hope that everything will turn out OK.

★ Fifth, we believe voters need to know what we've learned writing this book. This country is gearing up for the 2008 elections, and we hope you'll think about the information here when you listen to the candidates and decide how to vote. We don't have a recommended candidate, and we're not about to tell you that one party will handle the country's finances better than the other. So far, neither party has really stepped up to the plate on this issue, nor have the main presidential candidates put it front and center in their campaigns. Frankly, from where we stand, we don't expect much to happen unless and until voters begin to take the problem seriously—or until the country faces a financial crisis that would be extremely unpleasant for all of us. So it's time for voters to demand that candidates talk about this issue and spell out their approaches to it. It's time for voters to demand more candor and leadership from elected officials once they're in office. But to do this, voters need to be realistic themselves. There will always be politicians who promise more services and lower taxes—don't worry, be happy—but we don't have to buy their line.

★ Sixth, this issue changes all the time. *Where Does the Money Go?* includes a lot of information about what the government spends on this and that, which taxes collect how much money, and what experts predict for the budget if X or Y happens. Most of these numbers change every year depending on what Congress and the president do, the health of the economy, and other factors. As we take this book to press, there are a couple of important moving targets you need to keep an eye on. What Congress does or does not do about the alternative minimum tax and how long we keep large numbers of troops in Iraq are probably the major ones. Congress may make changes to the Medicare drug benefit that will either increase or dampen that program's costs to taxpayers. Changes like these are obviously important, and you do need to keep up with the news. But rest assured (if that's the right phrase in this instance), no matter what happens in Washington in the next couple of years, the big picture is the same. This country has a huge budget challenge coming up, and a couple of nips and tucks on taxes or spending will not make it disappear.

★ Seventh, it's taken us years to get ourselves in this mess, and no doubt it's going to take us years to get out of it.

There's just not a quick, simple way to turn this thing around. What's more, probably every single one of us will have to accept some changes we don't like. But keeping on the way we're going would be inexcusable. This problem can be solved. Let's get on with it.

CHAPTER 1

The Six Points You Need to Know to Understand the Federal Budget Crisis

Finance Minister: "Here is the Treasury Department's report, sir. I hope you'll find it clear."

Groucho Marx: "Clear? Huh. Why, a four-year-old child could understand this report . . . Run out and find me a four-year-old child—I can't make head or tail of it."

—Duck Soup, *1933*

Open any newspaper, tune in to any newscast, and someone will be tossing around billion- and trillion-dollar estimates about government spending and squabbling about the nation's finances. It certainly sounds important, but they don't make it easy for people who aren't policy wonks to understand. The numbers are mind-boggling, and the jargon is even worse. Unfunded liabilities, revenue neutral tax reform, entitlement spending, discretionary domestic programs, baseline assumptions, percentage of GDP. Faced with phrases like these, most of us reach for the remote to see what's going on in TVLand. But this debate is crucial to our future. Deep inside, you know it matters; otherwise, you wouldn't have opened this book.

The budget issue is a sneaky, slow-boil kind of a problem, one that's easy to avoid, and Americans have been doing just that for years. Politicians don't like to talk about cutting programs or raising taxes—which we'll no doubt need to do in some form or another in order to fix this budgetary mess. Journalists aren't making the country's budget problems the top news every night, either. After all, there are plenty of interesting scandals, crimes, and celebrity melodramas that make better headlines. And yes, fellow Americans, we've earned our share of the blame, too.

Let's be frank. When was the last time you cast your vote for a candidate who campaigned on getting the country's finances back on the right track? What about one who wants to cut government programs you like and raise taxes (which no one likes)? What do we talk about instead? Who had the most glittering celebrities at their fund-raiser. Who had the best zinger in the debate. Which candidate is making the cleverest use of YouTube. What a candidate did or did not do when he or she was twenty-something. No wonder so few people want to run for office these days—how would you like to have to defend everything you did and said in your twenties, or your thirties for that matter? And what about that pet question from the pre-pre-election polls: Which candidate would be more fun to have dinner with? How many Americans actually have dinner with presidential candidates, anyway? Go ahead, ask Mitt Romney or Barack Obama to meet you at Chili's sometime. See what happens. Is this what we really want from these people? Why are we spending time on this?

The truth is that those of us who aren't in government or

politics—those of us who generally watch from the sidelines trying to make sense of it all—had better start paying attention to the debate about the federal budget and the huge expenses we face in the coming decades. What's decided (or not decided) over the next few years will spell big changes for the way we live our daily lives. How the country solves or doesn't solve this problem will affect our paychecks, our investments, our mortgages, our kids' prospects in life, what kind of health care we'll get, our chances of ever getting to retire—even whether we live in a country that's fair, stable, and prosperous. And let's not kid ourselves. Right now, the savvy and well connected are already strolling the halls of Congress pushing for solutions that benefit them. So ignoring this debate is really not a very good option.

Fortunately, once you strip away all the confusing terms and unnecessary shouting, the budget problem isn't as hard to understand as the people in charge would like you to think.

THE BUDGET DEBATE, PARKING LOT VERSION

If you missed this on *Entertainment Tonight* or *Entourage*, a "parking lot version" is what Hollywood producer types call the shortest, simplest description of a movie or TV idea. Basically, it's what you can say to a studio exec if you're lucky enough to meet one in the parking lot, and you have to pitch your idea in the time it takes to walk to your cars. In Manhattan, which is short on parking lots, it's called the "elevator version."

We've reduced the budget issue to six essential points. Get these, and you're a long way to understanding what all the hoopla is about.

1. For thirty-one out of the last thirty-five years, the country has spent more on government programs and services than it has collected in taxes.

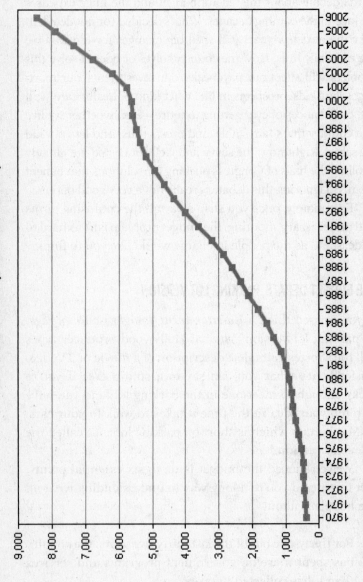

Gross National Debt (in billions of dollars)

The national debt always tends to increase, but as you can see, over the past few years the debt has leaped up, to the point where the U.S. government is now roughly $9 trillion in debt. *Source: Budget of the United States Government, FY 2008*

2. Every year the government comes up short, it borrows money to cover the difference. We've now built up a very big debt—roughly $9 trillion, and yes, that is *trillion* with a *t*.

3. The country will have humongous additional expenses over the next couple of decades as the baby boomers begin to retire and need more medical care.

4. There is no realistic way government can lower taxes (or even keep them at current levels), spend money on everything people want the government to do (at least according to the polls), and still end up with a balanced budget.

5. If we keep on going the way we're going, the debt will get bigger and begin to endanger the U.S. economy and our own personal finances and plans. And the government won't have enough money to pay for Social Security and Medicare for the boomers and still do what most of us expect government to do.

6. A substantial portion of the country's debt is held in foreign countries. Right now, these foreign investors consider U.S. government bonds one of the safest places in the world to put their money, but they could decide at some point that Europe or China or some other place is a better bet. This would be the global equivalent of a store clerk seizing your credit card and cutting it up.

If the country's state of financial affairs reminds you of people who spend more money than they make nearly every month, while cheerfully adding onto their credit card debt hoping that nothing will go wrong, you're not that far wrong. Obviously, government finances and family finances are different. For one thing, the government can probably raise taxes a lot more easily than most of us could just suddenly raise our own salaries substantially. The worst-case scenario is also much different. The president isn't going to walk out of the White House one morning and find out someone's repossessed the armor-plated limousine. But the concept is pretty similar. You can live beyond your means for quite a while without too much fuss (as long as nothing goes wrong). But at some point, the amount you owe begins to take its toll on the way you live.

SO WHY DO THEY KEEP DOING IT?

Given the dangers and the fact that everyone knows that the huge baby-boom generation is coming up for retirement now, *why* does the government keep on spending more than it takes in, you may ask in your best Cindy Lou Who voice. After all, politicians talk about balancing the budget and cutting the deficit all the time. The simple answer is there's always something people want more, whether it's tax cuts or better benefits or a stronger military. In some cases, that's a perfectly reasonable choice. During a recession, additional government spending can create jobs and rev up the economy, as Franklin Roosevelt did during the New Deal. The country also ran large deficits during World War II.

DEFICITS BEGIN TO FEEL NORMAL

There are good reasons for the country to run a deficit every so often, but if you do it long enough (thirty-one out of thirty-five years, for example), it starts to feel normal. It's nearly always easier for politicians to add to the debt than raise taxes or cut popular programs. In fact, in Washington these days, they don't even seem able to cut *unpopular* programs— if some politically connected someone somewhere likes it, it stays in the budget. And like a slow-growing tumor in the national economy, the debt just keeps on growing until it

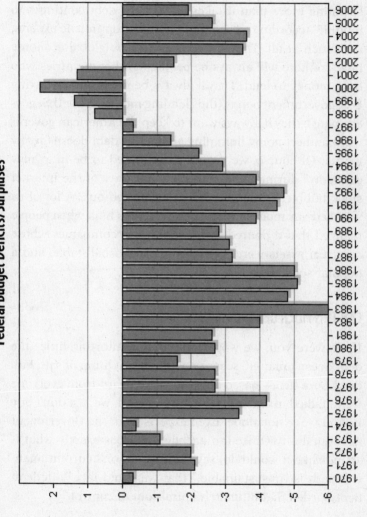

Federal Budget Deficits/Surpluses

Deficit spending has become routine for the U.S. government, with the government running in the red for thirty-one of the last thirty-five years. *Source: Budget of the United States Government, FY 2008*

becomes really dangerous—dangerous enough to affect the livelihood and lifestyle of nearly every single American.

What would actually happen if the country keeps on spending more than it takes in, and the debt continues to mount? According to one small group of optimistic experts, not much at all. They say we have a new global economy now, so there will always be people in other countries who have money to lend. They'll always be interested in buying U.S. government bonds (thus lending money to the U.S. government); they'll always want to keep the American government and economy humming along. The debt doesn't really matter. Of course, we were also supposed to be in a "new economy" during the lazy, hazy, crazy days of the Internet stock bubble. Remember how that turned out? A lot of us watched our modest little stock portfolios tank when people realized they'd poured their money into companies whose only real assets were a clever idea, a foosball table, and a popular sock puppet.

CAN THIS MUCH DEBT POSSIBLY BE OK?

If we were you, we would take that comforting little "the debt doesn't matter" scenario with a giant lump of salt. Full disclosure here—based on what we've heard from everyone we've talked to and everything we've read, we just don't buy it. And we're not alone. Even experts from the Government Accountability Office (an agency that does exactly what it sounds like it would do) say that the nation's current financial path is "unsustainable," that we could face "a federal debt burden that ultimately spirals out of control."[1]

---

[1] Government Accountability Office, "The Nation's Long-Term Fiscal Outlook April 2007 Update: The Bottom Line," GAO-07-983R (www .gao.gov/new.items/do7893r.pdf).

For us, it's much too dangerous to assume that being this much in debt is hunky-dory and that things will always go our way in the wide world of economics. We think it's just common sense and the better part of valor to assume that some of the risk is real. Risk is real. That's why we have auto insurance, home insurance, and health insurance and put money away for a rainy day.[2]

THREE REALLY BIG RISKS

So here's a sampling of what keeps a lot of economists and policy makers awake at night (well, it may not really keep them awake, but it does keep them writing articles and making speeches about how important it is to get the U.S. financial house in order).

No. 1: Monster expenses (that we know are coming) will wipe us out. Unless there's some sort of "Hot Zone" plague that wipes out a substantial portion of the baby boomers before they reach retirement (since your authors are boomers themselves, we're not recommending that), the government is facing gigantic increases in Social Security and Medicare expenses over the next couple of decades. This is the one big threat to the budget and the economy that nearly every reasonably sane expert we can find agrees on. Unless something changes, we could see a time (around 2040, if nothing is done) when nearly every tax dollar collected will be needed to pay for retirement and health care for the elderly and interest on the debt.[3] There will be almost no money for anything else, except maybe a basic national defense. You can read the grisly details in the coming pages.

---

[2] Well, that last one, we meant to do that. Lousy Internet bubble.

[3] Government Accountability Office, "The Nation's Long-Term Fiscal Outlook April 2007 Update."

No. 2: The economy goes down, down, down. When the government borrows more and more money, it makes it harder for the rest of us to borrow because it drives up interest rates. This risk is widely accepted among mainstream economists,[4] and it's pretty clear to anyone who has ever shopped for a mortgage why high interest rates are a bad thing. But mortgages are just the tip of the iceberg. High interest rates jeopardize nearly every part of the economy. Businesses can't get loans. Their costs go up. When they get in a pinch, they start cutting jobs. When the economy goes into a tailspin, there are more layoffs, fewer raises, more cuts in benefits, more businesses failing, bigger consumer debt, people's investments getting savaged, and more. Think "very, very bad recession."

No. 3: We'll come to regret relying on the kindness of strangers. About a quarter of U.S. government debt is held by foreign governments, banks, and investors, and right now, they seem reasonably happy to keep sending their extra money here. But the truth is that international politics and economics sometimes change in what seems to be the blink of an eye (remember back when you had never heard of Osama bin Laden?). What would happen if some big debt holders abroad suddenly wanted their money back right away? Would there be a mad scramble to raise taxes?

---

[4] It's basically supply and demand. The risk is that all this government borrowing will crowd out other lending. There's only so much money out there to borrow, and every dollar investors put into Treasury bonds is a dollar that isn't available for stocks, corporate bonds, real estate, venture capital, or anything else in the private sector. So the rest of us have to jump higher hurdles and pay higher interest to borrow money.

Would Congress suddenly slash spending with no time for the country to think about how to do it fairly? Would the stock market hurtle downward? Would American investors also lose confidence and start putting their money elsewhere? No one really knows. In fact, while most experts said this was a worrisome possibility, they didn't agree on exactly what would happen (comforting, isn't it?).

Some of these dangers could rise up suddenly—like the iceberg that appears dead ahead in that *Titanic* movie. People around the world who are happily buying U.S. Treasuries now could get anxious or disenchanted pretty quickly. Tech investors started dumping Internet stock with head-spinning speed back when that little stock market bubble burst. Or the changes could be gradual. We might face a slow decline in our standard of living, an economy that just never recovers, a government that is less and less able to provide services that people value. But note the recurring theme here: if things go badly, it nearly always comes out of *your* hide, as the taxpayer and citizen. Either life gets a lot more expensive, or you have to make do with less help from the government, or most likely both at once.

The good news is that there's still time to avoid this. It's like seeing a traffic accident a half mile down the road. You've still got time to slow down or change lanes. And we know this problem can be addressed, because the government was able to balance the books just a few years ago, thanks to a strong economy and some bipartisan financial realism.

The bad news is that politics as usual—the gridlock, the polarization, the sloganeering, the inability to compromise or move ahead on much of anything—could be setting the country up for a real smashup.

THE GREAT BUDGET DEBATE POP QUIZ

In this book, we'll try to avoid the wonk words and pie charts (well, we do have a pie chart or two). Even so, it's helpful to have a few facts in hand. So we'll start off with a pop quiz. If you've been watching C-SPAN and all those Sunday morning news shows, you'll probably ace this. However, if you're not as well informed, try the quiz anyway. You'll pick up a few things, proving once again that sleeping late on Sunday doesn't automatically keep you from getting ahead in life.

1. True or false? If it weren't for the war in Iraq, the federal budget would have been balanced the last couple of years.

ANSWER: *Not quite. The Iraq war has cost a lot of money, but the country wouldn't have balanced its budgets over the last few years even without the war (as amazing as that may seem). Through mid-2007, the country spent more than $400 billion on the war[5]—it's actually hard to tell precisely (more on that later). That's not pocket change to be sure, but during the same four years, the country added over $2.3 trillion to the debt (that's tril-lion with a t).[6] So it's not just the war—not by a long shot. You*

---

[5] Robert A. Sunshine, "Testimony on Estimated Costs of U.S. Operations in Iraq and Afghanistan and of Other Activities Related to the War on Terrorism," Congressional Budget Office, July 31, 2007 (www.cbo .gov/ftpdocs/84xx/doc8497/07-30-WarCosts_Testimony.pdf).

[6] Congressional Budget Office, Historical Budget Data, "Revenues, Outlays, Surpluses, Deficits, and Debt Held by the Public, 1962 to 2006 (www.cbo.gov/budget/historical.pdf).

★★★★★★★★★★★★★★★★★★★★★★★★★★★★★★★★

can say that the Iraq war made the already-hefty budget deficits for recent years bigger. But you can't say that without the war, the country would have been in the black. And remember, just balancing the budget isn't enough to solve the long-term problems with Social Security and Medicare. It's a good thing to do, no doubt, but it's only the beginning.

2. True or false? If we just rolled back the Bush tax cuts, we could solve all our problems with the federal budget.

ANSWER: *Nope, that's not true, either. President Bush and Congress enacted a series of different tax cuts—for families, investors, businesses—and most are set to expire at the end of 2010. In a democracy, it is much easier to give tax cuts than take them away, so hardly anyone thinks Congress will roll all of these taxes back to their Clinton-era rates. But even if that happened, it still wouldn't get us out of our financial jam. With the boomers beginning to retire, Social Security and Medicare costs are going to mushroom, and repealing the Bush tax cuts doesn't provide nearly enough money to cover the gap. All the major numbers crunchers agree on this one.[7] So you can be for 'em or agin 'em (much more on this in chapter 14), but you can't say that rolling back the Bush tax cuts will by itself solve the problem.*

3. Which of the following areas do budget experts worry about most—because its costs are so hard to control and could rise out of sight? A. The defense budget. B. The budget for FEMA, the agency that is supposed to help people when there are

---

[7] See, for example, Government Accountability Office, "The Nation's Long-Term Fiscal Outlook April 2007 Update: The Bottom Line," and testimony of Ben S. Bernanke, chairman of the Federal Reserve Board of Governors, before the House Committee on the Budget, February 28, 2007 (www.access.gpo.gov/congress/house/house04ch110.html).

★★★★★★★★★★★★★★★★★★★★★★★★★★★★

floods, hurricanes, and other big disasters. C. Social Security.
D. Medicare, which pays for health care for older people.

ANSWER: *D for Medicare. Its costs are skyrocketing out of sight. Right now, Medicare is 12 percent of the federal budget, and its costs are rising much faster than the rate of inflation. Since Medicare's costs are joined at the hip with the country's health care costs overall—that's what Medicare pays for, after all—predicting what will happen down the line is all the more complicated. Or as Federal Reserve chairman Ben Bernanke puts it, with the deadpan economist phrasing you have to have to run the Fed: "Projections of future medical costs are fraught with uncertainty."*[8]

4. True or false? Foreign aid is one of the top ten expenses in the federal budget.

ANSWER: *Not even close. Foreign aid is about 1 percent of federal spending each year.*

5. True or false? Money in the Social Security trust fund is only spent on Social Security.

ANSWER: *Nope. Money raised through Social Security taxes that is not immediately needed to pay Social Security benefits to elderly Americans can be lent (and has been lent) to the federal government to cover other programs and tax cuts. Sometimes people are distressed, even infuriated, when they learn about this, but the "borrowing" isn't illegal or even secret. In recent years, the huge baby-boom generation has been working and paying Social Security taxes, so there has been quite a bit of "extra money" in the trust fund. The rest of government has*

---

[8] Bernanke, testimony.

★★★★★★★★★★★★★★★★★★★★★★★★★★★★★

depended on the Social Security piggy bank to avoid having to raise taxes or cut other kinds of government spending. Social Security does have IOUs for money that's been spent elsewhere (actually, Treasury bonds), but it will need to start redeeming them when the boomers start retiring. Since the U.S. government routinely operates in the red and is now roughly $9 trillion in debt, coming up with the money to cover those IOUs is not going to be easy. See chapters 6 through 9 for the particulars.

Got all five right? Congratulations! You're ready to start your own "let's balance the budget" blog. If you missed something, read on. There's a lot to learn and think about in this strange little corner of public policy.

If you enjoyed this little exercise, you might want to check some of the other budget quizzes and games available online. For example, you can find out how much you know about how the government spends tax dollars by playing the "Online Penny Game" on the Concord Coalition's Web site[9] (Concord is a nonpartisan group focusing on budget issues).

---

[9] See www.concordcoalition.com.

So Who's in Worse Financial Shape, the U.S. Government Or Michael Jackson?

Credit: Library of Congress

Credit: Ronald Reagan Library

We know, this seems like a nonsensical question. The U.S. government is huge, spending unimaginable amounts of money and providing services that touch every American. The numbers are so colossal that budget experts sometimes resort to little *t*'s for trillions and *b*'s for billions so the zeros don't run off the page. Michael Jackson is wealthy, to be sure, but his entire fortune wouldn't keep the U.S. government running for even a day. Lots of people listen to his music and are fascinated by his life, but only his entourage is really depending on him. And while the income and expenditures of the U.S. government are public (you can get the entire federal budget at www.gpoaccess.gov/usbudget/), Michael Jackson's finances are essentially private. What we know, we know from news reports based on court filings. Still, the comparison offers some tidbits to chew on.

U.S. GOVERNMENT

Total Annual Income
$2.407 trillion (2006)

Total Annual Expenses
$2.655 trillion (2006)

Total Employees
2.7 million civilian employees
1.4 million active-duty military
personnel

**Wild, Bordering on
Inexplicable, Spending**
Has spent $640 each for
toilet seats and $792 on a

KING OF POP

Total Annual Income
Press accounts have estimated
Jackson's lifetime earnings
at $500 million, and he was
earning about $50 million a year
in the 1980s, but his income
has been going down steadily.[10]

Total Annual Expenses
$35 million (again, as estimated
in the press)[11]

Total Employees
He used to have a hundred or
so employees at Neverland
Ranch,[12] but now his entourage
seems to have dwindled. In
2006, the state of California
labor board temporarily closed
the ranch because Jackson
reportedly owed his employees
more than $300,000 in back
pay.[13]

**Wild, Bordering on
Inexplicable, Spending**
Reportedly once spent $10,000
on in-flight purchases on one

---

[10] Edna Gundersen, "For Jackson, Scandal Could Spell Financial Ruin,"
USA TODAY, updated November 25, 2003.

[11] Steve Chawkins and E. Scott Reckard, "Prosecutor Says Jackson Is on
the Brink of Bankruptcy," *Los Angeles Times,* March 12, 2005.

[12] Gary Strauss, "Jackson's Finances Are Solid, Adviser Says," *USA TODAY,*
April 28, 2004.

[13] *The Saturday Early Show,* Russ Mitchell, anchor, "Michael Jackson's
Neverland Ranch Shut Down by State of California," CBS News Transcripts,
March 11, 2006.

designer door mat.[14] "Pork barrel spending" is in the eye of the beholder, but you can always find examples like $50 million for an indoor rain forest in Iowa or $750,000 for grasshopper research in Alaska.

Swissair trip (granted, the in-flight movie can be pretty boring).[15] Contrary to creepy popular rumor, however, he never bought the Elephant Man's skeleton.

Unexpected Expenses
September 11, 2001
Hurricane Katrina

Unexpected Expenses
A criminal trial (at which he was acquitted), more civil suits than you can shake a stick at, not to mention fines for letting workers' compensation for his employees lapse.

Assets That Could Be Sold
National parks, the air traffic control system, prisons, public lands with oil, timber, and minerals, and a lot of planes and ships with low mileage. However, the public probably wouldn't stand for much of this.

Assets That Could Be Sold
He owns shares of a company with rights to music by the Beatles, Elvis Presley, Willie Nelson, and Pearl Jam, and he's been using them as collateral for loans.[16] As we write this, he still owns Neverland, but his Bengal tigers have been adopted by Tippi Hedren and his "people" are looking for homes for the other animals.[17] Frankly, we wouldn't be too surprised if both of these assets are gone before this book hits bookstores.

---

[14] Taxpayers for Common Sense, Senator William Proxmire's Golden Fleece Awards, 1985–1988, available at www.taxpayer.net/awards/goldenfleece/1985-1988.htm.

[15] Gundersen, "For Jackson, Scandal Could Spell Financial Ruin."

[16] Strauss, "Jackson's Finances Are Solid, Adviser Says."

[17] Michelle Caruso, "Former `Birds' Actress Adopts Michael Jackson's Tigers," *New York Daily News,* June 18, 2006.

Long-Term Prospects

You can't underestimate having the exclusive rights to print money and tax the world's largest economy. But even that may not be enough if we keep on borrowing and postponing decisions on how to cover the retirement and health care costs for the boomers. However, if we make some sensible decisions sooner rather than later, our prospects are pretty good.

Worst-Case Scenario

Crushing interest rates, steep tax increases, a tumbling stock market, a world financial crisis, and a government that can't do anything other than write Social Security checks and maintain an on-the-cheap national defense. Spending on everything else from national parks to student loans gets chucked.

Long-Term Prospects

Not nearly as good. King of Pop he may be, but not many showbiz pros see another *Thriller* in Jackson's future. His legal troubles have made him much less marketable. Unfortunately for Jackson, rumors that he was getting $10 million to attend a birthday party for the Sultan of Brunei's son turned out not to be true—or at least they were denied by the family.[18]

Worst-Case Scenario

The entourage is out of work, and the animals go to a zoo. Jackson might wind up as the Jackson family mooch with Janet and Jermaine earning bigger bucks. But even if he is totally bankrupt or winds up in jail, no more than a few hundred people will actually be affected—not an entire nation.

---

[18] "Scurrilous: Chicago's No. 1 Couch Potato," *Chicago Sun Times*, June 8, 2007.

CHAPTER 2

So What's the Worst That Could Happen?

Why would you want to be president in 2008? I don't understand it.

—Douglas Holtz-Eakin, former CBO director
and White House economic adviser

Experts can't predict exactly what will happen if the country keeps on adding debt and doesn't make changes in light of the upcoming retirement and health care costs of the boomers, but it isn't going to be good.

NOT ENOUGH NUMBERS ON THE DEBT CLOCK

Perhaps the most benign result will be that the national "debt clock" won't have enough digits to display the debt. Back in 1989, a real estate developer named Seymour Durst (he was worried about the nation's spending habits even then) paid to have a big sign put up in Midtown Manhattan that tallied the national debt pretty much like the mile-

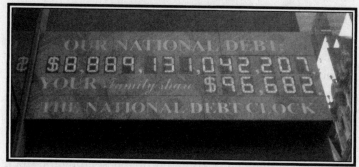

The late Seymour Durst spent his own money to put up the debt clock in the 1980s to call attention to the nation's debt problems. In September 2000, thinking the problem was under control after several years of federal surpluses, Durst's son Douglas turned the clock off. But it was a false alarm, and two years later Durst started the clock again. *Credit: Photo by Jean Johnson*

age odometer in a car. It also calculates the debt per family, approaching $100,000 in 2007. It was an imaginative idea, but Durst's imagination failed in one respect. Since the debt was less than $3 trillion back then—and given how unimaginably huge a trillion dollars is—the clock wasn't designed to show $10 trillion. It only goes up to $9,999,999,999,999.[1] Unfortunately for the clock and the country, we're at roughly $9 trillion and counting right now.[2]

So what will happen when the debt drifts up to $10 trillion and if it keeps going way beyond? In chapter 1, we pointed out three big risks. Here's a quick recap just in case you didn't take notes. Risk numero uno is that Social Security and Medicare for boomers will bust the federal budget wide open. To be precise about it, this one is not a "risk."

---

[1] U.S. Debt Clock Running Out of Time, Space, Yahoo News, AFP, March 27, 2006.

[2] Although it looks to us like they could use the space where the dollar sign is now for the extra digit when the debt rolls over to $10 trillion. We expect that everyone will still get the point.

It's a certainty unless we make some adjustments in the way these programs work and are paid for. Risk number two is that the country's colossal debt and all the borrowing needed to handle it begin to drag the economy down. We could have the lovely symptoms of an economic recession around for a very long time. And risk number three is that people holding billions of dollars in U.S. debt—say, in China—will suddenly decide to put their money elsewhere, unleashing a financial crisis and probably sending the stock market into the basement.

HITTING THE WALL

A financial crisis is a lot like dying. For some people it happens slowly, like cancer. For others it's fast, like a car crash. The debt could hit us either way.

Let's say we (meaning the United States) keep going as we are. We don't rein in spending. We don't raise taxes. We keep Social Security and Medicare just the way they are now (maybe we even increase Medicare benefits because people are so worried about health care costs). Remember that handy little GAO (Government Accountability Office) calculation? Unless we make some changes, by 2040, nearly every dollar the government collects in taxes will be needed for Social Security, Medicare, and paying interest on the debt.[3] Obviously, the government wouldn't be able to do a lot of things we normally expect—invest in research, offer college loans, enforce health and safety laws, take care of the national parks, and so on. And let's say that the government, lacking the guts to do anything that might upset vot-

---

[3] Government Accountability Office, "The Nation's Long-Term Fiscal Outlook April 2007 Update: The Bottom Line," GAO-07-983R (www .gao.gov/new.items/d07983r.pdf).

ers, keeps borrowing money to pay the bills. And the rest of the world keeps buying our Treasury bills because we're still the safest place for foreigners to park their money.

Until something happens.

The "something" could be a lot of different things. Something could seriously disrupt the world's oil supply, for example—a war in the Mideast or a wave of Islamist revolution that brings hostile new regimes to power. Or something could undermine the U.S. economy dramatically, such as the collapse of a major industry or the failure of an overstrained power grid. You can sit around all day and easily come up with your own list of possible disasters if you enjoy doing that sort of thing.

It could be something slower, too. Perhaps the U.S. economy just starts looking a little less competitive compared with markets overseas, a little sluggish, not quite keeping up with other nations on the science-and-technology front. Not a crisis, but something bankers and bond traders talk about until they reach one of Malcolm Gladwell's famous "tipping points."

SOME BANKER IN BEIJING

But either way, it starts like this: a banker in one of those glass-walled office buildings that look the same whether they're in Beijing, London, or Singapore rubs his eyes, walks over to the vending machine, thinks a bit, sits back down, and says to his colleagues, "You know, I think we're holding too many U.S. Treasuries."

After years of having U.S. government bonds be the money market fund to the world, it's the awful moment. The idea spreads. Washington can't sell the bonds anymore. The Treasury Department jacks up the interest rates to make them more attractive. When the Treasury rate goes up, all the commercial banks raise their rates, too. Everything from your credit card bill to your mortgage gets more expensive overnight. And the foreign banks still aren't lending us enough money to cover the government's costs.

Now Washington doesn't just have a deficit problem, it's got a cash flow problem. And it's got to be solved now. Elected officials start raising taxes and cutting government programs in an attempt to come up with the cash and convince foreign investors that the country means business about getting its economic house in order. The economy starts to slump because of the high interest rates and the sudden shock of huge tax hikes. The stock market plummets as people lose confidence and draw on their stocks to try to maintain their lifestyle. Even American investors start moving money overseas. How many people are really patriotic enough to stand by and watch the savings of a lifetime lose value?

Still hard to visualize? It's time to bring all this a little closer to home. What would all this mean for a typical American family, people who work for a living, have kids, try to save for the future, hope to be able to retire at some point? Now we could make up an imaginary family, but why should we when Frank Capra has already done most of the work? We're going to borrow a very-well-known family living in that classic all-American hometown, Bedford Falls.

BEDFORD FALLS IN A DOWN-AND-OUT AMERICA

If you're reading this in December, we probably don't have to refresh your memory about Bedford Falls. It's Jimmy Stewart's hometown in the 1946 Christmas movie *It's a Wonderful Life*. If you're one of those people who absolutely refuse to watch it, Jimmy Stewart is good-guy George Bailey, whose family runs the local savings and loan while the

Things always turn out all right in the end for George Bailey and family in the Christmas reruns. But Clarence the angel is no financial planner, and we may all be living in Pottersville unless we put the nation's fiscal house in order. *Credit:* It's a Wonderful LIfe, *1946*

villainous Mr. Potter runs the local bank. Thanks to Mr. Potter's nefarious dealing, Bailey gets into financial trouble, and thinks about killing himself. He wishes aloud that he had never been born.

In what's often viewed as the movie's signature sequence, guardian angel Clarence escorts a "nonexistent" George Bailey around Bedford Falls so he can see what the town would have been like if he had in fact never been born. Instead of being a Norman Rockwell vision of comfortable American middle-class life, Bedford Falls is now Pottersville. The wicked Mr. Potter has been given free rein, and Bailey's family, friends, and his hometown generally have hit upon hard times and various misfortunes.

In the movie, Bedford Falls becomes Pottersville because

one man (Mr. Potter) is greedy and awful. But Bedford Falls could also become Debtville (sorry, it was irresistible) because millions of Americans aren't paying attention to the country's budget problems and because too many of their leaders apparently lack the courage, candor, and integrity needed to tackle them before there's real trouble.

Let's take a look at what happened in Pottersville without George Bailey and what might happen in Debtville if the country's addiction to spending beyond its means keeps on going.

POTTERSVILLE

Mr. Potter takes over the Bailey Building & Loan, and people end up living in Potter's own run-down development with the somewhat alarming name of Potter's Field.[4] George Bailey wasn't there to help people of middle-class means buy nicer, but affordable homes.

Mrs. Bailey, George's mother, has become a bitter old woman forced to take lodgers into the family home to make ends meet. George wasn't there to make sure she lived in reasonable comfort.

Uncle Billy, George's bungling but lovable uncle, is in an insane asylum and, Mrs. Bailey reports, no one has seen him for years. George wasn't there to protect and guard over him.

Mary, George's wife, is a spinster working at the local library. She and George never fell in love and never married.

Zuzu, Bailey's daughter, doesn't exist because her dad wasn't around to marry Mary and start a family.

---

[4] That man needed a good publicist more than anyone in movie history.

Harry, George's war hero brother, also doesn't exist. He drowned as a child because George wasn't there to save him.

Bert and Ernie, George's old buddies, the local cop and taxi driver, are gruff, rough, and sour men, not the generous, funny, friendly characters who inspired the *Sesame Street* Muppet names.

DEBTVILLE

In Debtville, **Mr. Potter** doesn't have to do the dastardly deeds himself. Massive federal debt drives up interest rates pushing the country into recession. Chinese investors decide they want their money back (and other investors around the world follow suit), setting off an economic crisis. The stock market loses a quarter of its value in less than a month. In a last-ditch attempt to pay the Chinese and cover basic expenses, Congress raises taxes and slashes support for education, small businesses, local police, housing, and other services. Even Social Security benefits take a hit. The financial situation is so grave that Congress even ends the politically popular home mortgage deduction. With high interest rates and no deduction to ease the pain, owning a home becomes a dream for most middle-class Americans.

Still, characters like Mr. Potter thrive even in the worst economic times. At the first hint of trouble in the U.S. economy, he moved his money to the Cayman Islands. His rundown Potter's Field rental complex is bursting at the seams.

Mrs. Bailey, George's mother, is still bitter and taking in lodgers. When the debt crisis hit the government, her Social Security benefits were cut, and the little nest egg she had in the stock market nearly evaporated. Mrs. Bailey is barely making ends meet, and she's worried about getting sick. Medicare, too, has been slashed.

Uncle Billy used to get supplemental Social Security payments because of his disabilities, which enabled him to live decently and independently. But these funds were gutted in the budget emergency. Like many of the country's most vulnerable people, Uncle Billy has joined the ranks of the homeless and the hungry (food stamps were cut back, too; they're just for families now—not enough money to help singles).

Mary still works at the library, but since it's only open a few hours a day, she's having a tough time, too. The library was funded by the state and the town of Pottersville, but with all the federal cutbacks, local governments have had to scramble to make up the difference. Keeping the library open on a full schedule is too much of a luxury.

We've let **Zuzu** live (maybe she's someone else's daughter), but she isn't going to college. With the government spending nearly everything it takes in to cover even pared-down Social Security and Medicare benefits, college loans are a thing of the past. She's living at home with her mom—jobs are few and far between for people just coming into the workforce.

We've let **Harry** survive, too, and he's a war hero like he was when George Bailey lived his normal life. But despite Harry's service to the country in time of war, his veterans' benefits were cut. Normally, this is one of the last places the American public wants to scrimp, but the government's financial crisis was so sudden and so dire, even the most broadly supported government programs have been lacerated.

Bert and Ernie? Well, Bert the police officer is working overtime and without much backup. The poor economy and high rates of unemployment have caused crime rates to soar. And the Pottersville Police Department has had to take its share of cutbacks, just like the library did. As for Ernie the taxi driver, the government's double-digit increases in the gas tax, and the fact that hardly anyone can afford to take a cab these days, have devastated his livelihood as well.

That's the thing about a government in crisis and an economy in a deep recession. Nearly everyone suffers, except for the very rich.

When Good Economies Go Bad, or, Don't Cry for Me, Argentina

What happens when a government hits the wall and can't pay its creditors? Ask Argentina. In December 2001, there were riots in the streets over Argentina's financial problems, riots so bad that martial law was declared and two presidents were forced from office. Argentina's situation is very different from our own, but what occurred there should get our attention about what happens when a country can't pay its bills.

Like the United States, Argentina is rich in natural and human resources, but economically speaking, the country has had its ups and downs. In the late 1980s, Argentina was facing "hyperinflation" of 200 percent *a month*—the kind of inflation where it pays to go grocery shopping in the morning because by nightfall things will cost more. To fight this, Argentina made several policy changes encouraged by the International Monetary Fund, including pegging the peso to the U.S. dollar and liberalizing trade rules. Argentina also started borrowing heavily from the IMF and other international banks, hoping to keep its budget going until the reforms took hold.

For a while in the 1990s, that plan worked—in fact, it worked really well. Argentina saw strong economic growth and started getting cited as an economic model for others. But when the world market for farm products softened and currencies in other Latin American countries fell, Argentine products had trouble finding a market. That led to a recession, which cut tax revenues to the government.

The IMF and international banks told Argentine officials to go on a strict government austerity plan, but with unemployment at 18 percent they (not surprisingly) resisted that until 2001. Then the government devalued the peso and slashed its budget dramatically, including a move to cut old-age pensions and use the money to pay international debts.

The country's credit rating sank and there was a national run on banks, with Argentines pulling $1.3 billion out of their accounts in a single day. Dozens were killed in street protests so serious that the government imposed martial law, froze bank deposits to stop the run, then defaulted on more than $141 billion in foreign debt—the biggest loan default in history. The recession that followed was brutal. More than half of the population was in poverty between 2001 and 2002.[5] Unemployment hit nearly 21 percent and the economy contracted by 11 percent.[6] By contrast, in the 1982 U.S. recession, the most severe in recent years, the economy shrank 1.9 percent and unemployment reached 9.6 percent.[7]

Some six years later, Argentina is working its way out of this corner. The government has paid off its IMF debt early, after an elaborate debt-swap deal that left its international creditors get-

[5] International Monetary Fund Country Report 05/236, July 18, 2005.

[6] "IMF Executive Board Concludes 2006 Article IV Consultation with Argentina," International Monetary Fund Public Information Notice No. 06/93, August 9, 2006 (www.imf.org/external/np/sec/pn/2006/pn0693.htm).

[7] Bureau of Labor Statistics, "Employment Status of the Civilian Noninstitutional Population, 1940 to Date," accessed June 16, 2007; U.S. Bureau of Economic Analysis, National Income and Product Accounts Table, accessed June 16, 2007 (www.bea.gov/bea/dn/nipaweb/TableView.asp#Mid).

ting 35 cents on the dollar.[8] Unemployment has fallen, and the economy is growing at a healthy 8 percent. But Argentina's overall debt still totals nearly 90 percent of its gross domestic product. The burden to average Argentines has been immense (tens of thousands have reportedly left the country to find work and more than a quarter are below the poverty line).[9]

The United States isn't Argentina. Even if the worst happens here, it probably won't happen in the same way. The U.S. economy is much bigger and more diverse than Argentina's, plus we have the advantage that the U.S. dollar is the standard currency for international banking. Argentina had to borrow (and repay) its debt in U.S. dollars, which was pretty tough when the value of their own currency, the peso, was in free fall. You can also have a really passionate argument on whether the IMF's advice led Argentina into a blind alley. That probably won't be a factor for the United States.

But the point to remember is that the national debt—that abstraction that lives on the computers of big banks—became very real to the Argentine people. The debt cost them real cash and real jobs because their government mishandled it. And that absolutely could happen to us.

---

[8] BBC News, "Argentine Restructuring 'Success,'" March 4, 2005.

[9] Mark P. Sullivan, "Argentina: Political and Economic Conditions and U.S. Relations," Congressional Research Service, October 12, 2006.

So Where in the World Is the Debt?

Today, the debt owed by the United States is roughly $9 trillion, and as of 2007, more than $2 trillion of that amount is owed to foreign banks and other international investors. This is a nice vote of confidence in the U.S. economy in some respects. People around the world think their money will be safe invested in U.S. Treasury bonds. Still, there's a risk that at some point international investors might decide to put some of their money elsewhere, and that could drive up interest rates and cause turmoil in the stock and currency markets—not such a pretty picture.

The Top Ten Foreign Holders of U.S. Debt

Country	Amount
Japan	$612.3 billion
China, Mainland	$420.2 billion
United Kingdom (including the Channel Islands and the Isle of Man)	$145.1 billion
Oil exporters (including Ecuador, Venezuela, Indonesia, Bahrain, Iran, Iraq, Kuwait, Oman, Qatar, Saudi Arabia, the United Arab Emirates, Algeria, Gabon, Libya, and Nigeria)	$113.0 billion
Caribbean banking centers (includes the Bahamas, Bermuda, Cayman Islands, Netherlands Antilles, Panama, and British Virgin Islands)	$84.4 billion
Brazil	$70.6 billion
Luxembourg	$61.6 billion
Hong Kong	$58.7 billion
Korea	$58.1 billion
Taiwan	$57.9 billion

Source: Department of the Treasury/Federal Reserve Board, May 15, 2007

The China Syndrome, or How Bankers in Beijing Affect Your Mortgage

So when the U.S government borrows money, who does it borrow from? Well, anyone who's willing to buy Treasury bonds, for a start. Lots of Americans—individuals and institutions—own U.S. Treasuries. So do lots of foreigners. And in the last few years, few nations have had a bigger appetite for U.S. Treasury bonds than the People's Bank of China, Beijing's central bank.

Right now, China's economy is growing fast, and it's a country that exports a lot more than it imports (mainly because low-wage labor makes Chinese goods cheap to produce, but doesn't give Chinese workers enough spending money to acquire a taste for imported products). That means China is taking in a lot more money than it can spend, and it needs to stash it somewhere. China's central bankers, being cautious as bankers usually are, put that extra cash into safe investments. And Treasury bonds have always been among the safest investments going.

As of 2007, China owned $420 billion in Treasury bonds and is steadily buying more. That provides a lot of benefits to the United States—and poses a lot of risk.

In addition to allowing the U.S. government to go on spending money it doesn't have, the fact that we can always count on China to buy T-bills helps keep U.S. interest rates down. When somebody's always willing to lend you money on good terms, nobody has to haggle and rates stay low for everybody. That means your mortgage payments, car payments, and credit card bills are lower than they might be if the U.S. government had to struggle to get anyone to take its bonds.

The risk is pretty simple: Maybe someday the People's Bank of China will decide it doesn't want Treasury bonds anymore.

Maybe it will decide something else is a better investment. Or maybe we'll get into an argument with the Chinese over the future of Taiwan or human rights and China will get ticked off enough to start unloading its bonds on the world market. That could drop the dollar through the floor in international markets, not to mention jacking up every adjustable rate mortgage in the United States overnight. All of a sudden, you could find yourself with a lot less spending money every month.

The Chinese don't even have to stop buying bonds completely to cause us a lot of trouble. All China needs to do is slow down to set off alarms among currency traders around the world. Everybody in the financial world is watching for a sign that China has changed its mind, so they'll know when to dump their Treasury bonds, too.

Ironically, this is what makes U.S. debt a mixed blessing from the Chinese perspective. China needs stable world financial markets, and it needs Americans to keep buying Chinese products. All of which means the Chinese don't want to shake up the U.S. economy too badly. So they can't stop buying bonds, even if they could make more money elsewhere. It's another version of the old adage that "if you borrow $100,000 the bank owns you, but if you borrow $100 million you own the bank."

So does that mean there's no real risk? We're happily codependent forever? Well, maybe—or maybe not.

In the short term, what this most likely means is that China doesn't really have to listen to the United States on issues like human rights. Ever borrow money from someone and then try to give them advice or, worse, lecture them on morality? No, neither have we, but we don't think it would work. Do we really want to be in that position with the Chinese government? One in which American diplomats are trying to push an issue and the Chinese officials are sitting there thinking, "Yadda yadda yadda, wonder how many T-bills we bought today?"

In the long run, it's also dangerous to assume China will need our gizmo-hungry consumers forever. Someday Chinese consumers will take up some of the slack, or people in new emerging markets will get rich enough to start buying their own gizmos, and then maybe China won't need the United States so much anymore. Plus, there may be flashpoint situations (like Taiwanese independence) in which Beijing might be willing to sacrifice its own prosperity to get what it wants.

Look, the Japanese and the British hold a lot of Treasury bonds, too, but nobody makes a big deal about it. The Bank of England has no political reason to stick it to us. (Although it's worth remembering that if U.S. finances get screwed up badly enough, even our close allies might find other places to park their money. As the Corleones like to say in *The Godfather* movies, "It's not personal. It's strictly business.")

How you feel about this depends on what you think about the Chinese government and what it's after. China has a powerful weapon it could use against us, which it would rather not use. But because China is a rival instead of an ally, we can't assume it never will.

CHAPTER 3

A Little Clarification Is in Order

There are a lot of things people don't understand. Take the Einstein theory. Take taxes. Take love. Do you understand them? Neither do I. But they exist. They happen.

—*Screenwriter Dalton Trumbo in*
The Remarkable Andrew, *1941*

Most books that try to explain complicated topics—topics most of us don't chat about daily with friends and family—include a glossary of the hard words to make things clearer. You've lucked out here because we've decided to spare you that. After all, it's the key facts—and the decisions the country needs to make—that we want you to think about, not the specialized words economists and politicians use.

THE D-WORDS

Even so, there are two words we do need to spend some time on. Reporters and politicians tend to fling them around like crazy, and unfortunately, they both start with *d,* just to add to the potential confusion. They are, however, very different things, and making sure that you don't mistake **the deficit** for **the debt** is important to understanding this issue. (OK, you're thinking, "Oh come on, everyone knows that." Excellent; you've been paying attention while some of us were a little distracted. You move on to chapter 4.) For the rest of us, let's just get this little mix-up out of the way.

First, there's **the deficit.** When the government spends more money in a year than it collects in taxes and fees, it has a deficit for that year. In 2005, the U.S. government took in about $318 billion less than it spent, so it had a $318 billion deficit for that year. In 2006, the government took in about $248 billion less than it spent. So for that year, the deficit was _____? Yes, you've got it—$248 billion.

Then there's **the debt,** which is what the government owes when deficits add up over time. You can see that the deficits for 2005 and 2006 alone add up to really big bucks (2005's $318 billion + 2006's $248 billion = $566 billion). And that's just two years' worth. So after spending more money than it collects for thirty-one out of the last thirty-five years, by the close of 2007 the United States had built up a debt of roughly $9 trillion. Just pause on that for a moment. The country's debt is about $9 *trillion*, with a *t*— not $9 billion.

The dilemma for those of us with things to do and people to meet is that deficits go up and down all the time. Unless the country actually has a surplus, the debt is really just getting bigger. For example, based on what happened with the 2005 and 2006 federal budgets, you could write a nice headline that says, "2006 U.S. Deficit Down by $70 Billion." Sounds terrific, and it's even true. But it's also true that the government just added more than $500 billion to the debt (remember—2005's $318 billion deficit added to 2006's $248 billion deficit for a whopping $566 billion).

HERMIONE'S DEFICIT SPENDING FALLS IN JANUARY

Dealing with numbers in the billions and trillions and using words like *deficit* and *debt* can make the whole thing confusing, but the basic financial dilemma is common enough. Plenty of American households are in the same situation. Let's take our friend Hermione. She has a pretty good job, but month after month, she uses her credit cards to live beyond her means—a new leather sofa here, that new designer coat there. Now she owes more than $22,000 on her credit cards. Some months are a little better than others, of course. In December (we all know how this goes), Hermione added $1,000 to her credit card debt. In January (after making some New Year's resolutions in this area), she added only $100 to the overall total. You could write an entirely truthful headline about Hermione's little step in the right direction: "Hermione's January Deficit Spending Falls by $900." But Hermione is still in big financial trouble. She overspends. She underpays. And despite her New Year's resolutions, she's still courting financial trouble.

So the next time you see a news report about the federal deficit dropping by billions of dollars, don't get too excited. It's certainly better than the deficit growing, but it hardly means that the country is home free. For one thing, the federal deficit tends to be bigger or smaller depending on the state of the economy, regardless of whether the government has done anything to get its spending in line with its income or not. When the economy is good, people and companies make more money, and they pay more in taxes. During the economic boom between 1998 and 2001, the government actually ran a surplus for a few years. The opposite is also true. When the country is in a recession, people and companies earn less and pay less in taxes, and the deficits tend to be bigger.

What's more, virtually all economists believe there are times when the government should run a deficit. Sometimes it is just sensible and necessary. For example, the U.S. government ran large deficits during the Great Depression and throughout World War II. After the war, the government's record on deficits was mixed. Some years, the country had a surplus; some years, it had a deficit.[1]

HEADING IN THE WRONG DIRECTION

Since the 1970s, however, the country has been indulging in a steady diet of deficits. We did have those nice four years of surplus between 1998 and 2001, but the overall picture has not been pretty for quite some time. By 2007, the accumulated debt reached about $9 trillion, a mind-boggling and totally meaningless number for most of us.

It's probably fair to say that most economists, politicians, and business leaders are beginning to worry about that number, but there are some who believe it's not especially terrible in an economy as large and powerful as ours. After all, our friend Hermione's $22,000 credit card debt may be excessive for her, but Bill Gates or Oprah Winfrey wouldn't even blink if they had to send in a check for the entire amount.

MORTGAGING THE FUTURE?

So maybe the country's having a big debt is just like a family having a big mortgage? Well, yes and no. Right now, the country's debt ($9 trillion) is over three and a half times the

---

[1] "Federal Debt at End of Year, 1940–2012," Historical Tables, Budget of the United States Government, Fiscal Year 2008, accessed May 29, 2007 (www.gpoaccess.gov/usbudget/fy08/pdf/hist.pdf).

federal government's annual income from taxes and fees (about $2.4 trillion in 2006)—along the lines of a person making $50,000 a year carrying a $175,000 mortgage. Like those of us with mortgages, the government also pays some hefty interest on its debt, but for now, at least, the country's economy seems to be able to handle it.

But there are some major differences. People with mortgages follow a very specific plan to pay off what they owe. The U.S. government has no such plan. Only a few members of Congress are even talking about developing one. Most people with mortgages don't routinely add more and more to their debt every year—and those who do take on second mortgages, big home equity loans, and lots of credit card bills typically end up in trouble. In contrast, the government routinely adds to the debt. And finally, if you're a homeowner with a mortgage, and things start getting financially treacherous, you can always sell your home and go back to renting. You may not get as much as you hoped—sometimes you even have to take a loss—but in most cases you can sell your property and use the proceeds to pay down your debts.

Theoretically, the U.S. government could sell off everything it owns (condominiums in Yellowstone, anyone?). State governments do that pretty regularly—selling off some piece of infrastructure to raise cash, say, an office building or a turnpike, and then leasing it back. On paper, the U.S. government's total assets, including facilities and inventory, total $1.4 trillion.[2] But in the real world, that's just not a plan most Americans will tolerate. Even state governments that do this frequently end up getting criticized for phony

---

[2] Government Accountability Office, "Fiscal Stewardship: A Critical Challenge Facing Our Nation," January 2007 (www.gao.gov/new .items/d07362sp.pdf).

accounting. Ever hear those news reports about a state or city with a miserable bond rating?

THE REAL KICKER

But the most nerve-racking thing about the government's "let's add to the debt and worry about it tomorrow" mentality is what's to come. The country is quickly approaching a time when it will face some very big new expenses. Leaders in government and business all know this, and you should, too, because, as a country, we need to decide how to handle it. Members of the baby-boom generation are starting to retire. Very soon, they'll begin to have the kind of high health care expenses older people typically have. The government will have to pay for most of these expenses (trust us on this for now; we'll explain in chapters 6 thought 9). Unfortunately, the country is doing almost nothing to prepare for this.

This is the big financial crunch to come—the one that could mean serious problems for our economy and our way of life. To tackle that one, federal deficits need to fall this year and next year and keep on falling for a good number of years. We need to watch our expenses like the perennial hawk. And we're probably going to need to cut some expenses in ways we would rather not. It's like our friend Hermione, who needs to stop buying things unless she really, really needs them. And she needs to start paying more than the minimum on her credit card bill for many months to come. And she needs to drop that expensive gym membership and start running in the park instead.

The rest of *Where Does the Money Go?* will help you think for yourself about how the country should address this problem. We'll explain how the government gets its money, how much it takes in, what it spends it on, what

Social Security and Medicare have to do with it, and what kinds of choices there are to address the problem. We aim to give you information that will help you begin to gather your own thoughts about how to put things in balance. And we aim to give you the facts that will help you—good citizen that you are—start holding our elected officials' collective feet to the fire on this.

WHEN A BILLION DOLLARS DOESN'T MEAN MUCH

But before we go there, here's one last friendly reminder about understanding the budget debate. We all know this, but with all the numbers and statistics flying off the shelves these days, we just have to remember to watch those billions and trillions. For you or me, a billion dollars is a ton of money. For the federal government, it's chicken feed. Remember, in its 2006 budget, the government spent $248 *billion* more than it took in. And the country's debt is estimated at about $9 *trillion*.

The point is this: Just don't be too impressed when you hear or read that the government's deficit is down by $30, $40, or even $50 billion. That's not going to do the trick, especially if it's just for one year. As a country, we need to decide whether we want to cut back on government spending, change the way Social Security and Medicare work, raise taxes, or do some of all three. And after that, we're really going to have to stick with the plan.

What's a Billion Worth?

The trouble with big numbers is that they're hard to visualize. The $60 you take out of the ATM on Monday morning is crisp and tangible. You know what it takes to get through the week, and you know whether you're going to have to stop at the bank again before Friday.

A billion is just a number. There's no billion-dollar bill (although it's fun to speculate which president would be on it).[3] A trillion is even worse. These are important ways of keeping score, but really difficult to grasp. And when you start talking about big numbers, you ought to know how much they really mean in practice—as Dr. Evil found out as he tried to look threatening when he made his demand for "one *million* dollars."

But the best way of dealing with intangibles is to make them concrete. There's a Barenaked Ladies song called "If I Had a Million Dollars." Like some other songs, it's (a) been co-opted for a TV commercial and (b) can be difficult to get out of your head if you're not careful. But if you had a billion dollars and an inclination to play Santa, you could:[4]

★ Buy about 200 million bottles of aspirin (or about 143 million bottles, if you go with a name brand. See how it pays to buy generic?)

A billion dollars buys 200 million bottles of aspirin.
Credit: IStockphoto.

---

[3] Ever wonder why certain famous people are on particular bills—for example, why Hamilton's on the $10 and Jackson's on the $20, instead of the other way around? Well, the official answer from the Treasury Department is "We don't remember." The current lineup was set in 1928 and the records don't cover why. See www.ustreas.gov/education/faq/currency/portraits.html.

[4] Our figures are based on retail prices in August 2007—not counting taxes, coupons, or bulk purchase discounts.

★ Give five spiral notebooks to every student in public school in the United States (about 33.5 million children)

★ Get a pair of Gap jeans for everyone in Australia (20 million, not counting shipping). Or, if you'd rather work closer to home, that's enough to give all 5 million people in Minnesota their own iPod nano.

A billion dollars buys five spiral notebooks for every U.S. public school student. *Credit: Photo by the U.S. Census Bureau, Public Information Office (PIO)*

A billion dollars buys nearly 2 million Manolo Blahniks. *Credit: IStockphoto.com*

★ If you're a shoe-obsessed fashionista like Carrie in *Sex and the City,* you could buy yourself 2 million pairs of Manolo Blahniks. (If you're shoe-obsessed but less self-absorbed, that would allow you to give a pair to everyone in Manhattan and Staten Island. Including the men.)

★ You could also provide all thirty-five thousand students at the University of Houston with their own Toyota Camry. Not the base model, either—

A billion dollars would buy a fairly nice car for each of the University of Houston's thirty-five thousand students. *Credit: Photo by the U.S. Census Bureau, Public Information Office (PIO)* you could get the optional leather seats and high-end sound system.

★ You could keep about forty-five thousand people in a four-year private college for a year—or, depending on their behavior, in prison. The College Board says private tuition and fees average $22,218 a year; the Bureau of Justice Statistics says the average cost per inmate is $22,650 a year.

★ You could pay the salaries of about eighteen thousand rookie cops to put on the streets of Los Angeles (starting pay: $54,475 in 2007). There are only about ninety-two hundred officers in the LAPD now.

★ You'd have enough money to build a thousand-bed hospital (the rule of thumb is that a hospital costs $1 million per bed). Or, if you already had the hospital, you could conduct cardiac valve replacement surgery on nearly twenty-three thousand patients.

★ Based on the Census Bureau's median prices, you could buy homes for nearly 12,700 families in Mississippi, or nearly 3,158 homes in pricier California.

Big-ticket items, of course, cost more, and you get less for your money. A billion will only get you:

A billion dollars buys 3,158 big, fancy houses in California. *Credit: Photo by the U.S. Census Bureau, Public Information Office (PIO)*

★ Four Boeing 777 airliners (the extended-range model, for those long vacations).

★ Almost half of a new *Virginia*-class nuclear submarine, at $2.3 billion each.

★ One-third of New York's proposed Freedom Tower ($3 billion).

But what about trillions? They're really mind-boggling. To figure that out, just add three more zeros to any of the numbers above.

For example, instead of building a thousand hospital beds with $1 billion, $1 trillion would allow you to build a million hospital beds—and in 2005, there were only about 840,000 in the United States.

A billion dollars would pay for about half of a nuclear submarine. *Credit: U.S. Navy photo by Gene Royer*

Or instead of only four Boeing 777s, you could get 4,000, enough to replace the fleets of six or seven major airlines (for example, American Airlines and its American Eagle subsidiary have about 980 planes).

But let's make it as down-to-earth as possible. There are about 300 million people in the United States, more or less. And let's say you wanted to do something for every one of them. With $1 billion, split evenly, you'd have a little more than $3.33 to spend on each one.

With $1 trillion, you could spend $3,333.33 apiece.

Pay No Attention to That Headline

We don't usually recommend ignoring the news, but headlines saying the budget deficit is falling can be monumentally misleading. The **deficit** shows whether the government has balanced the budget in any year. If there's a deficit big or small, the budget is not balanced. The **debt** shows how much money the government has borrowed in total.

The big message here? Don't let politicians and the chattering classes gloat just because the deficit is smaller than it was last year. That's good, but it's not good enough. And remember, just balancing the budget won't solve the problem. If we don't tackle the long-term financial problems with Social Security and Medicare, the country will still find itself in budgetary hot water.

Here are some headlines that look pretty good, but the country's debt was rising relentlessly to its record $9 trillion level.

JANUARY 27, 1990: "THE 1991 BUDGET: BUSH'S PRO-POSAL; BUSH BUDGET SEES GROWING REVENUE AND CUT IN DEFICIT"[5]
SEPTEMBER 30, 1990: NATIONAL DEBT IS $3.2 TRILLION

NOVEMBER 22, 1993: "OCTOBER DEFICIT **DOWN** 7.1 PERCENT"[6]
SEPTEMBER 30, 1993: NATIONAL DEBT HITS $4.4 TRILLION

MARCH 20, 1997: "AS BUDGET BATTLES RAGE, THE DEFICIT IS SHRINKING ANYWAY"[7]
SEPTEMBER 30, 1997: NATIONAL DEBT HITS $5.4 TRILLION

---

[5] Robert Pear, "The 1991 Budget: Bush's Proposal; Bush Budget Sees Growing Revenue and Cut in Deficit," *New York Times,* January 26, 1990.

[6] John D. McClain, "October Deficit Down 7.1 Percent," Associated Press, November 22, 1993.

[7] John M. Berry, "As Budget Battles Rage, the Deficit Is Shrinking Anyway," *Washington Post,* March 20, 1997.

OCTOBER 24, 2000: "US GOV'T POSTS RECORD 2000 SURPLUS"[8]
SEPTEMBER 30, 2000: NATIONAL DEBT HITS $5.6 TRILLION

JULY 22, 2005: "GOOD NEWS ON THE DEFICIT"[9]
SEPTEMBER 30, 2005: NATIONAL DEBT HITS $7.9 TRILLION

JANUARY 25, 2007: "CONGRESSIONAL OFFICE FORECASTS DROP IN DEFICIT, WITH POSSIBILITY OF A LATER SURPLUS"[10]

SEPTEMBER 30, 2007: NATIONAL DEBT REACHES $9 TRILLION

★

A Deficit Here, a Deficit There, Here a Deficit, There a Deficit . . .

You might as well brace yourselves because you're going to read the word *deficit* a few hundred times in this book. Here, we're talking about the federal budget deficit—the gap between what the U.S. government spends on programs and services and what it takes in from taxes and a few other sources. Obviously, it's an important issue because when the government runs a deficit—which it is doing routinely—the country's debt and the interest we

---

[8] Jeannine A. Versa, "US Gov't Posts Record 2000 Surplus," Associated Press, October 24, 2000.

[9] Mike Rosen, "Good News on the Deficit," *Rocky Mountain News* (Denver), July 22, 2005.

[10] Edmund L. Andrews, "Congressional Office Forecasts Drop in Deficit, with Possibility of a Later Surplus," *New York Times*, January 25, 2007.

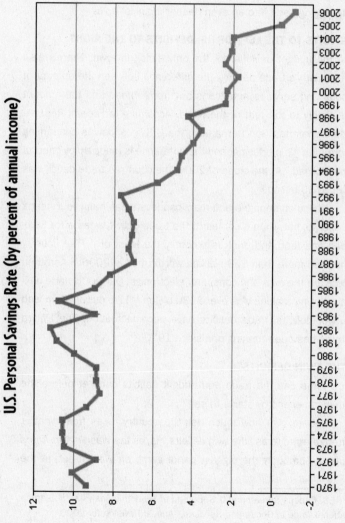

U.S. Personal Savings Rate (by percent of annual income)

Sometime over the last thirty years or so, Americans just stopped saving money, in good times and in bad. The savings rate has plummeted, and that has real implications for the nation's economy and for the federal budget. *Source: U.S. Bureau of Economic Analysis* pay on it swells. Rather than saving up to cover the big expenses ahead when the baby boomers stop working, the country is actually digging itself into an even deeper financial hole.

DEFICITS TO THE LEFT OF US, DEFICITS TO THE RIGHT

But the budget deficit's not the only deficit in town. There's also the country's trade deficit—the difference between the amount of goods and services Americans buy from abroad and the amount we supply to the rest of the world. According to Federal Reserve Board chairman Ben Bernanke, "the U.S. economy is consuming more than it's producing, and the difference is made up by imports from abroad." At the close of 2006, the country's trade deficit was over $750 billion.[11]

All you champagne and marzipan lovers can hang your heads in shame, but when it comes to the trade deficit, a taste for European fine food and drink is probably the least of it. The country imported more than $299 billion worth of oil in 2006.[12] And then there are the cars and consumer electronics. Our 2006 trade deficit with China alone was over $230 billion.[13] Like our problem with budget deficits, trade deficits have become routine. The United States has been running one since 1970.

"THE TWIN DEFICITS"

How large can the trade and budget deficits can get before the American economy starts to sag?

Debate over how much risk the country faces from what is often referred to as "the twin deficits," is, as the *Washington Times* puts it, "probably the biggest parlor game on Wall Street, at the

---

[11] U.S. Census Bureau and U.S. Bureau of Economic Analysis, "U.S. International Trade in Goods and Services," Annual Revision for 2006.

[12] Ibid.

[13] Martin Crutsinger, "Government Won't Cite China on Currency," Associated Press Online, June 13, 2007.

Federal Reserve and inside think tanks and university economics departments in recent years."[14] It's a complicated question involving the value of the dollar, interest rates, and the degree to which investors around the world see the United States as a good place to stash money. Most economists believe prolonged and expanding trade deficits are signs of a troubled economy, but others say the trade issue isn't dire yet given the wealth of the U.S. overall. Harvard economist Robert Lawrence is in the latter group. "We are so rich as a country," he told the *Christian Science Monitor*. "We're borrowing, we're running down our assets, but we're very wealthy."[15]

A SCARY FUTURE?

But that's not all, as they say in all those infomercials selling knives and folding colanders. The United States is also the proud holder of yet a third deficit—the savings deficit. Since average Americans don't sock away a lot of money in the bank, and since we continue to buy all those alluring new products and rack up credit card debt, we're not helping out, either. As a group, American consumers were actually in the red in 2005 and 2006.[16] Add the savings deficit to the budget and trade deficits, and you begin to get an unsettling economic picture. Our government spends more than it takes in, so it needs to borrow. Our economy imports more than we export, so we need money to cover that gap. Since Americans don't save and invest all that much (unlike China, where people are savings whizzes), the country needs to borrow money from abroad to keep

---

[14] "Is the Trade Deficit Sustainable?" *Washington Times,* December 29, 2005.

[15] Mark Trumbull, "Giant Trade Gap; No End in Sight," *Christian Science Monitor,* March 10, 2006.

[16] Personal Savings Rate, National Economic Accounts, U.S. Bureau of Economic Analysis, www.bea.gov/briefrm/saving.htm, accessed June 17, 2007.

the whole thing glued together. So far, so good—but can we really keep this up?

U.S. comptroller general David Walker is one of the few government officials to talk forcefully and repeatedly about the risks the country is running with its free-spending, freewheeling financial ways. Walker often refers to four deficits—the budget deficit, the trade deficit, the savings deficit, and the leadership deficit. "We are in much worse financial condition than advertised," he says. "The future is scary."[17]

---

[17] Rob Christensen, "Doomed by Debt?" *News & Observer* (Raleigh, N.C.), March 19, 2006.

LAYING DOWN THE BASELINE

Here you'll see three different projections about what's likely to happen to the federal budget over the next few years. And as you can see, two of the three don't look that bad, with the budget going into surplus around 2012.

How can the numbers be so different? Well, there is a reason.

There are plenty of organizations in the budget-projection business, but for lots of people in Washington, the rule has been "when in doubt, use the CBO figure." The Congressional Budget Office is an independent, nonpartisan agency without a cause to promote. They're all about the numbers. You'll see that the CBO

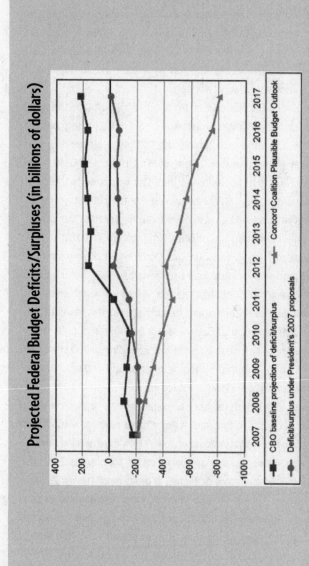

Projected Federal Budget Deficits/Surpluses (in billions of dollars)

Legend:
- CBO baseline projection of deficit/surplus
- Deficit/surplus under President's 2007 proposals
- Concord Coalition Plausible Budget Outlook

When it comes to projecting the federal budget deficit, it all depends on the assumptions you make. The Congressional Budget Office, the White House Office of Management and Budget, and the Concord Coalition end up with very different estimates about the deficit over the next decade. *Source: Congressional Budget Office, Office of Management and Budget, and the Concord Coalition* projects that the deficit will actually decline and change to a small surplus by 2012. That's assuming, however, that all of the Bush tax cuts will expire on schedule in 2010, producing a surge of revenue. Not likely, by nearly all accounts. So if they're so smart and independent, why did they say it? Because by law, the CBO is required to assume that the law as written will stand, even if everyone knows that won't happen.

The second projection is from the White House Office of Management and Budget, which is also full of very sharp budget experts, just like the CBO. Unlike the CBO, they all work for one human being: the president. That doesn't mean the OMB numbers are wrong, but it does mean that it's committed to the president's agenda. The OMB is assuming the tax cuts stay and its estimates of economic growth have been higher than the CBO's. But they also estimate the deficit falling to zero in 2012.

The third is from the Concord Coalition, a prominent nonpartisan group advocating fiscal responsibility—in other words, "deficit hawks." In Concord's figures, it makes some guesses about political reality, including assuming that the tax cuts will be made permanent, that Congress will limit the unpopular alternative minimum tax, that government spending will exceed the rate of inflation, and that spending on Iraq and Afghanistan will come down. Its projections are a lot grimmer.

Our advice? Whenever you see people or organizations making estimates about the federal deficit going up or down, ask what their assumptions are. Do taxes go up or down? How much will the government spend? Does the economy grow, and if so, how fast? Unless you know these basics, any projection is pretty much meaningless.

Despite all their disagreements about projections in the short term, however, CBO, OMB, and Concord all use the same term when they look at the long-term outlook: "unsustainable." So whatever the budget looks like over the next few years, in the long run none of the projections look good.

★★★★★★★★★★★★★★★★★★★★★★★★★★★

CHAPTER 4

The Tax Tour
(or Money Comes...)

Death and taxes and childbirth! There's never any convenient
time for any of them!

—*Scarlett O'Hara in* Gone with the Wind,
Margaret Mitchell, 1936

In some respects, the budget issue is fairly simple. After all,
there are only a couple of ways to look at it. By this point, we
think you'll agree that ignoring the problem and hoping it will
go away is not an option. (If you're still set on that strategy,
there's got to be something worth watching on ESPN.) Other
than that, you're left with raising taxes or cutting spending or
putting together a plan where you do some of both.

There are experts who think the country can make a
good dent in the problem by cutting waste in government
and keeping the economy growing at a healthy pace (when
the economy is good, the government collects more taxes
because people are earning more money). We discuss these
strategies and what they could do for the budget picture in
chapters 10 and 14. But almost no one we can find thinks

there is enough sheer waste in government to offset the deficits the country is running now—not to mention the red ink we're facing with the big baby-boom expenses coming up. And while it would be great to have an economy that boomed all the time, it's not clear that anyone really knows how to make that happen—especially not in the rapidly changing and confusing world economy we have now.

Sources of Federal Revenue, 2006

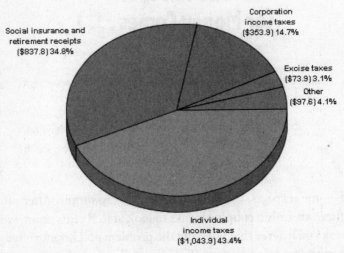

Personal income taxes—the money withheld from your paycheck and the forms you submit every April 15—are the largest single share of federal revenue, bringing in nearly 44 percent of the government's money in 2006. *Source: Budget of the United States Government, FY 2008*

TIME TO GRAB THE ASPIRIN

Given that, we have to think about raising taxes or cutting spending or finding some "not too cold, not too hot" combination platter. So now it's time to take a closer look at the options, starting with the tax tour. You might think of it as "Taxes 101." It's designed to get you thinking about taxes in general. We return to the tax question and the big choices facing us in 2010 in chapter 14.

Some of you may want to grab the aspirin right now, because for a lot of Americans, the very thought of taxes produces a headache. Partly it's those three-inch-thick books on display in your local bookstore every January—the ones on how to prepare your own tax return. Or the fact that millions of not particularly rich people feel they have to buy software or hire someone to do their taxes for them (nearly everyone agrees the U.S. tax code is insanely complicated). And partly it's that fear of having to make a check out to the "U.S. Treasury" on April 15. Taxes are just not a cheery topic.

Plus, it's only fair to warn you right up front: we're entering territory where experts don't agree and where, just in case you've been asleep at the wheel for the last decade, there's a lot of political fistfighting. In the end, you're going to have to make your own decision based on conflicting "facts" and competing claims. But honestly, that's not a big deal. If you've signed up for cell phone service lately, or bought a flat-screen TV, you're used to that. We make decisions based on contradictory, confusing information all the time.

So here goes. The U.S. government gets money, which totaled $2.4 trillion in 2006, from four main sources.

INCOME TAXES

Nearly half of the money the government spends comes from individual income taxes paid by you and me. For nearly everyone reading this, that includes people who are poorer than you and people who are richer (Mr. Gates, if you're reading this, you're the exception). Income tax rates have fallen fairly dramatically since World War II, when the top tax bracket was 94 percent, meaning that if you earned over

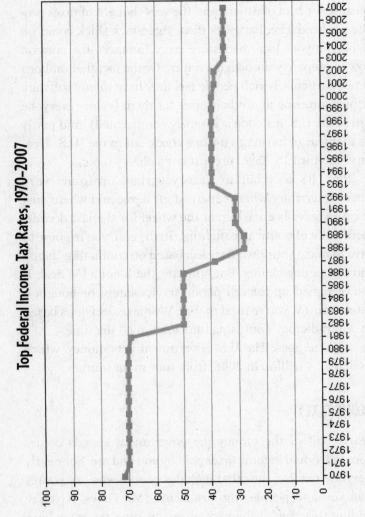

Top Federal Income Tax Rates, 1970–2007

The very top income tax rate—the so-called marginal rate—has fallen dramatically since the early 1970s.

Source: Tax Policy Center

$200,000 (which was a lot of money back then), the government took 94 percent of what you received over that.[1] Those rates dipped down over time, but by the time President Reagan came into office in 1981 the top rate was 70 percent. The Reagan tax cuts brought the top bracket down to 50 percent. In the 2000 presidential election, President Bush campaigned vigorously on his plan to cut taxes, and in 2001 and 2003, Congress passed big tax cuts. The top tax bracket is now 35 percent.[2] Some 18 million Americans earn so little that they don't even need to file a tax return, while another 33 million end up paying nothing at all.[3] With certain tax credits for children for example, some receive money back even though they are too poor to owe taxes. For a variety of political reasons, nearly all of the tax changes enacted under President Bush are set to expire in 2010, so the battle over whether to keep them is already under way. As most liberals see it, the cuts are a major reason the country has had such big budget deficits the last few years. Most liberals don't want the tax cuts extended—at least those affecting higher-income Americans. As most conservatives see it, the tax cuts have helped the economy and let people keep more of their own money. For conservatives, too much government spending is what's causing the budget problems. You can read a lot more about what's at stake in the big tax battle ahead of us in chapter 14.

---

[1] This is what economists and accountants call a "marginal tax rate"—it's the tax rate you pay on the last dollar you earn.

[2] Tax Policy Center, Individual Income Tax Brackets, 1945–2007, www.taxpolicycenter.org/TaxFacts/Tfdb/Content/PDF/individual_rates .pdf, accessed May 20, 2007.

[3] Tax Policy Center, Nonfilers and Filers With Modest Tax Liabilities, 2003, http://www.taxpolicycenter.org/TaxFacts/TFDB/TFTemplate.cfm? Docid=283, accessed May 20, 2007.

Payroll Taxes

About a third of the money the government spends comes from Social Security and Medicare taxes, unemployment taxes, and retirement payments made by federal employees. Social Security and Medicare take nice little chunks out of your take-home pay—those dreaded FICA deductions on your pay stub, although Social Security taxes stop for people who earn over a certain amount ($94,200 in 2007). Because of this, Social Security taxes are often considered "regressive" since they tend to hit nearly everyone on the lower end of the income scale while letting the people who earn the big bucks off the hook at a certain point. One consolation is that your employer has to pay, too (the same amount as you), and, in theory at least, you'll get some personal benefit from these taxes down the road when you retire or need health care as an older person. The money from the payroll tax you and your employer pay is used to cover Social Security and Medicare for people who are elderly right now. These programs were designed that way—Social Security back in 1935 and Medicare in 1965—so they could help older people who needed it right away (chapters 6 through 9 have a lot more on this). Because the baby-boom generation is so large, and because most boomers are still working and paying these taxes, there is a surplus in the Social Security program. However, the government "borrows" from this surplus to cover other kinds of expenses, and so it won't have to raise income taxes (more about this in chapters 6 through 9 as well).

Corporate Taxes

About 10 percent of the money the government spends comes from taxes on corporations. You're probably thinking ExxonMobil or Time Warner when we say *corporation*, but this category also includes little shops and restaurants, dry cleaners, hair salons, exterminators, and other small businesses owned by families and individuals. If you have any 401(k) money invested in stocks, you can count yourself as something of a business "owner," because whatever happens with corporate taxes affects you, too. Because so many of us picture "big business" when the subject of corporate taxes comes up, we often assume these taxes could easily be raised without much impact on us. After all, these corporate dudes work in awfully nice offices, wear really nice clothes (there's a reason why they're called "suits"), and you always seem to see them flying business or first class. There's no reason why big corporations shouldn't be asked to pay their share of the nation's tax burden, but there are complications to keep in mind. One is that we want corporations to hire people and give them good salaries and benefits, and the more taxes businesses pay the less money they have for their workers. We also want corporations to invest in technology and new products so we don't get eaten alive in the global marketplace by all these up-and-coming foreign economies. And finally, we want these companies to stay here, not to move to some other country where labor is cheap and the government is less ambitious. So when we're looking around for people to pay more taxes, the idea of socking it to business is a little more complicated than it initially might seem.

Alcohol, Tobacco, and Gas—the Excise Taxes

About 3 percent of the money the government spends comes from excise taxes, mainly on alcohol, tobacco, airline tickets, and gas. Three percent might seem skimpy given how steep these taxes are when you have to pay them. Actually, that little 3 percent was about $74 billion in 2006, a huge amount of money even though it's a small piece of the government's overall income. But there's another reason these taxes generate comparatively small amounts for the federal government even though they seem pretty hefty when you fill 'er up or buy a bottle of Johnnie Walker. Some of this money goes to state and local government. This varies by state, of course, and it's one of the reasons why smokers in New York and New Jersey think about stocking up on cigarettes if they visit Mississippi or Missouri.[4]

There are a few other smaller sources for the money the government spends, but we'll leave those for you to check out yourself if you're curious. We've listed several useful Web sites in the appendix.

HOW MUCH BLAME DOES WASHINGTON DESERVE?

This last point about the excise taxes on gas and cigarettes brings up another issue that sometimes gets lost in the hub-bub. The feds are hardly the only tax collectors around. We pay property taxes to the states and towns where we live (it's included in the rent even if you don't own a home yourself). Most of us also pay state and local sales taxes. Many states have their own income tax. Some states tax new-car sales and real estate transactions. Some interesting calculations by Kevin Hassett, a senior fellow at the American Enterprise Institute, suggest that for those worried about the tax burden on middle- and lower-income families, state and local taxes are the real culprit—not the federal income tax. "That's because the federal income tax, which is steeply progressive—the higher your income, the more you pay in taxes—gets all the media attention. But other taxes that are less visible, such as sales taxes, hit lower-income families with a heavy thud," Hassett writes.[5] So while it's convenient

---

[4] "State Excise Tax Rates on Cigarettes," Federation of Tax Administrators, available at www.taxadmin.org/fta/rate/cigarett.html.

[5] Kevin Hassett, "Why We Pay without a Whimper," *Washington Post,* April 15, 2007.

to blame "Washington" when we think about taxes, remember that it's not Washington alone.

But whether it's federal taxes or those collected by states and cities, views on whether to raise taxes to deal with budget problems often depend largely on whether we believe government programs and services are useful and helpful. Essentially, we ask ourselves whether what we're getting is worth paying for. In the next few chapters, we'll try to help you understand where the government spends money now, how much is "wasted," and whether some of what government does now could be done better and less expensively by someone else—maybe even you personally.

But before we close out our tax primer, there are a few other questions to think about. Just looking at the arithmetic, there's little question that raising taxes would help the country's budget's bottom line. If more money comes in, there's less of a hole. But before you leap, spend a moment or two considering these four questions. The experts are all over the map on these, so your own cogitating on these may not be so bad.

DO TAX CUTS PAY FOR THEMSELVES?

Some prominent leaders—actually, very prominent ones like President Bush and Vice President Cheney—point out that in fact tax cuts can bring money into the U.S. Treasury. For Vice President Cheney, lower taxes are "a powerful driver of investment, growth, and new jobs for America's workers. And that increased economic activity, in turn, generates revenue for the federal government."[6] The argument goes like this: When the economy grows, people

---

[6] Vice President Dick Cheney, speech to the Conservative Political Action Conference, Omni Shoreham Hotel, Washington, D.C., March 1, 2007.

and businesses make more money. When people and businesses make more money, they pay higher taxes. And when
those higher taxes come in, they help cover any shortfall
caused by cutting taxes in the first place. Those advancing
this line of argument often point to what has happened in
the last several years. In 2001 and 2003, Congress passed
large tax cuts, and recently revenues to the U.S. Treasury
have indeed increased. According to the Treasury Department, government revenues are up 35 percent since 2003
because of "solid economic growth and improved corporate
tax yields."[7] Some people and companies have made a lot
more money.

But is it enough money to cover the cost of the tax cuts?
Well, unfortunately, no. Not even conservative experts—those
who generally want taxes to be low and tend to back President Bush's economic policies—think tax cuts actually bring
in enough extra money to pay for themselves. Economists'
estimates vary depending on which taxes you're talking about
and what kind of assumptions and economic projections their
calculations make. It's not a simple matter predicting exactly
what will happen. But well-respected economists like N. Gregory Mankiw, who chaired President Bush's Council of Economic Advisers, and Douglas Holtz-Eakin, who worked in the
Bush White House and for Congress, both have said that tax
cuts don't bring in quite that much money. Mankiw estimated
that cuts in capital gains taxes (paid on profits from selling
property or stocks) generate revenue to cover about half their
costs; Holtz-Eakin put the "replacement value" for cutting personal taxes up to 22 percent for the first five years and up to
32 percent in the following five.[8] The *Washington Post*'s Sebas-

[7] 2006 Financial Report of the United States Government, Department of the Treasury, Executive Summary, p. 4.

[8] Douglas Holtz-Eakin, Congressional Budget Office, "Analyzing the

tian Mallaby, who covered this controversy in his column, concludes that the "free-lunch mantra is just plain wrong."[9]

Of course, the argument doesn't really stop there, and it probably shouldn't. Most knowledgeable experts agree that specific tax cuts don't generate sufficient revenue to pay for themselves. However, there is an important debate about whether an economy with very low taxes works better over the long term than an economy with higher ones. So that brings us to our next question.

WILL RAISING TAXES HARM THE ECONOMY?

It would be nice to give you a simple yes-no answer on this one, but we couldn't find it. Most economists are worried about the country's budget problems, especially the gargantuan financial hole we face when the boomers start to retire and need health care in big numbers. And there's no doubt that at some point raising taxes too much can harm the economy. This is because money that could be invested (which is good for the economy) or spent on products and services (also good for the economy) goes instead to the government.

Raising too many taxes too quickly can upset the stock market and lead investors to look for opportunities in other countries instead of here. It can sap the enthusiasm of entrepreneurs, inventors, and others who take risks and put their money into new economic ventures. Offering these risk-

Economic and Budgetary Effects of a 10 Percent Cut in Income Tax Rates," December 1, 2005, and Gregory Mankiw's analysis at http://economics.harvard.edu/faculty/mankiw/files/dynamicscoring_05-1212.pdf.

[9] Sebastian Mallaby, "The Return of Voodoo Economics: Republicans Ignore Their Experts on the Cost of Tax Cuts," *Washington Post*, May 15, 2006, A17. See also editorial, "A Heckuva Claim; Mr. Bush Is Oblivious to the Consequences of His Tax Cuts," *Washington Post*, January 7, 2007.

taking folk plenty of incentive to do their thing has given the United States a pretty good economic run for a very long time now, so you do have to be careful about killing the goose that lays the golden egg. Raising taxes on very specific parts of the economy can also pack an unexpected wallop. See "Lost at Sea: A Short History of Taxing Yachts," on page 73, to find out what went wrong when Congress decided that taxing very big, expensive boats would affect only the very rich.

On the other side of it, federal income taxes were significantly higher in the 1950s and 1960s and the U.S. economy had plenty of very good years back then. Robert Rubin, one of President Clinton's economic advisers (and secretary of the treasury from 1998 through 1999) argued that raising taxes to reduce the deficit could help the economy grow. After his recommended policies took hold ("Rubinomics" it was called), the economy perked up very nicely for a number of years. Some economists and policy makers point out that when the government uses tax money to invest in education, research, highways, air traffic control, and other services, this can help the economy. There's also the argument that the government itself buys goods and services, which is also good for the economy. Just ask anyone who lives near a military base.

Others like economist and *New York Times* columnist Paul Krugman question whether strong economic growth, spurred by low taxes, really means anything if the benefits aren't shared broadly. An ardent critic of the recent Bush tax cuts, Krugman acknowledges that the economy has grown, but says that the main result has been a growing gap between the very wealthy and other Americans. "Where did all the economic growth go?" Krugman asks. "It went to a relative handful of people at the top."[10]

---

[10] Paul Krugman, "The Great Wealth Transfer," *Rolling Stone*, November 30, 2006.

Tax Burden in Selected Countries (by percent of income)

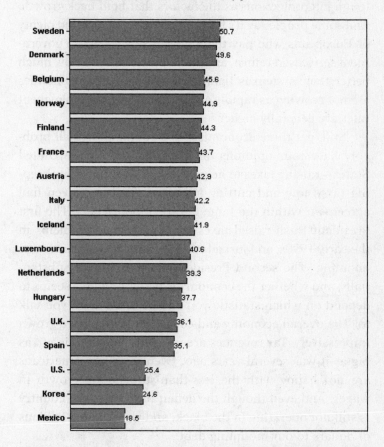

Country	Percent
Sweden	50.7
Denmark	49.6
Belgium	45.6
Norway	44.9
Finland	44.3
France	43.7
Austria	42.9
Italy	42.2
Iceland	41.9
Luxembourg	40.6
Netherlands	39.3
Hungary	37.7
U.K.	36.1
Spain	35.1
U.S.	25.4
Korea	24.6
Mexico	18.5

The total tax burden in the United States (including state and local taxes) is considerably lower than in most European countries. *Source: Organization for Economic Cooperation and Development*

Some experts believe that Europe—where taxes are quite a bit higher than here—is at or near the point where high taxes eat away at and weaken the economy, and that high unemployment rates there show this; others point to the Europeans' longer vacations, shorter workweeks, and tough job protections as the factors that hold back growth. And some people say it's both. Of course, you can find plenty of Europeans who pay the higher taxes, get more government services in return, and are happy to tell you how much better their system is than ours—even if their economies are not growing as rapidly as ours and their unemployment rates are generally higher.

Still, get three economists in a room, and you'll probably have four opinions on what's better for the United States—raising taxes to help get the budget in line or keeping taxes low and cutting spending. You can't even find agreement within the same family of politicians. The first President Bush raised taxes to solve budget problems in the early 1990s, and just a few years later the economy was booming. The second President Bush cut taxes substantially, and whether the economy is better or worse seems to depend on which statistic you choose and whom you talk to. The overall economy and corporate profits have grown impressively. Tax revenues are up, and the deficit is not as big as it was several years ago. Even so, many Americans are not happy with the less-than-stupendous growth in wages. And even though the deficit has fallen, the country is still not operating in the black, and we've added trillions of dollars to our mounting debt.

WILL IT BE FAIR?

Another big wrinkle in the tax issue is whether, if we raise taxes, we can do it in ways that are fair. We all bring our own values to the table on this one. Conservatives typically believe that individuals should be able to keep most of what they earn, and that government's call on our tax dollars should be very, very limited. They don't think government does that many things all that well; they say individuals make better decisions about where money should go. Liberals typically believe that government has done many good things with taxes (Social Security is a favorite example), and that wealthier Americans should help pay for services that help and protect the poor and the middle class. The wealthy, they argue, have benefited from our society, and there's nothing wrong with asking them to pay for things like health care, child care, and college loans—things that give other people a chance. Besides, most liberals say, the wealthy will barely even miss it.

It's just two different points of view. Still, the situation gets confusing when advocates for both sides start throwing in the dueling statistics. Consider the argument over the income tax cuts ushered in by President Bush. As most conservatives tell it, these tax cuts benefited nearly all Americans. As most liberals tell it, most of the cuts have gone to the wealthy. Actually, both of these statements are true. How does it work? We'll use an imaginary illustration because examples from the real-life U.S. tax code with all its brackets and deductions are nearly incomprehensible.

Suppose the government cuts taxes by 10 percent. Someone who owes a million dollars in taxes saves $100,000, enough for a couple of Mercedes, or, if you prefer, Hummers. But someone who owes $1,000 saves $100—nice to have, certainly, but more in the Razor Scooter price range. So both the millionaire and the typical Joe benefited. But most of the tax cut went to the millionaire. Is it fair—allowing people to keep more of their own money? They made it, after all. Or is it just a windfall for the rich—one that should be ended so we can get the budget back on track?

There's an old Washington saying (attributed to the late Louisiana Senator Russell Long) that boils tax policy down to three short phrases: "Don't tax you, don't tax me, tax that fellow behind the tree." Hardly anyone wants to pay taxes, and most of us think there's someone else somewhere who should be paying more, so we don't have to pay as much. In the end, it's probably impossible to create a tax system that meets everyone's vision of fairness.

DOES IT HAVE TO BE SO HARD?

And the fourth big complication is the complication itself. If you've ever done your own taxes, you don't need to be told that the federal tax code is fiendishly dense and convoluted. A quick trip to the post office or the IRS Web site for tax forms will confirm that. There's also reason to believe it's getting worse, not better. Federal tax rules and regulations have increased from 40,500 pages in 1995 to more than 66,000 pages in 2006[11]— *War and Peace* is a pamphlet by comparison. By one analysis, businesses, nonprofits, and individuals spent 6 billion hours calculating their taxes and otherwise complying with federal tax rules, at a total cost of $265.1 billion.[12]

So does anyone have any good ideas for making things simpler? Sure, but it's surprising how complicated simplicity can be. The problem is that if you want to keep the basic structure of the progressive income tax we have now (rather than move to a national sales tax or a flat tax, which would be simpler plans, whatever you might think of them), the job's not as easy as it sounds. A progressive income tax system—where higher-income people pay a higher tax rate—is by its very nature more complicated. After all, you have to keep track of what people are earning, a big challenge in

---

[11] Cato Institute, "The Simple (Tax) Life," April 17, 2006, available at www.cato.org/pub_display.php?pub_id=6345.

[12] Tax Foundation "The Rising Cost of Complying with the Federal Income Tax," January 10, 2006, available at www.taxfoundation.org/publications/show/1281.html.

and of itself. That's what led to our existing system, which withholds taxes from your paycheck before you even see it, and in which banks and investment firms have to report your interest and dividends directly to the IRS.

Another hurdle is that most exemptions and exceptions in the tax code—the features that make it so intricate and complex—are there to help and please potential voters. You're probably already grousing about big business and big oil, and if you are, you're not alone. But in all likelihood, you're a big beneficiary, too. Some of the features that make current tax forms so complicated are also very popular and aimed at average people: deductions for home mortgage interest and charitable contributions; getting a tax break to save for college or retirement, or to pay for child care.

And that brings up a related issue. Although the main purpose of the tax system is to raise revenue for the government, we also use it as a tool for social policy—to encourage people to do good things and discourage others. If we want businesses and individuals to buy more fuel-efficient vehicles and energy-efficient appliances, we give them tax breaks for doing it. If we want people to save more for college or retirement, we add tax breaks for that, too. You're able to deduct your mortgage interest because, as a society, we've decided it's better for everyone if people own their own homes rather than rent. The tax code would be simpler if we didn't do this, but since getting out of paying some tax is such a good motivator, our society has found this strategy very useful. If we give up the idea of using taxes for social change, just how *are* we going to do these things?

Nearly everyone—except maybe tax lawyers and accountants—likes the *idea* of simplifying taxes, and politicians often say that this is what they, too, want. Scott McClellan, President Bush's press secretary between 2003 and 2005, once claimed that the president's effort to repeal the estate tax was making the tax code simpler. The president "is open to ideas that move us in the direction of a simpler and fairer tax code," McClellan said. "And one thing that—one real important step we took to make the tax code simpler was to eliminate the death tax. We need to make that permanent. That is a great way to simplify the tax code; you eliminate 90 pages in the tax code right there."[13] Of course, that still leaves more than 65,000 pages of tax regulations, instructions, and clarifications. And since most Americans never have to worry about the estate tax anyway, this may not be quite what most are hoping for.

In chapter 15, we describe the short, bleak life of the President's Advisory Panel on Federal Tax Reform, a bipartisan commission put together by President Bush to look at how to simplify taxes. The panel's ideas haven't gotten much traction to date, but you might want to see what the panel suggested to ease what they call the "headache of burdensome record-keeping, lengthy instructions, and complicated schedules, worksheets, and forms—often requiring multiple computations that are neither logical nor intuitive." You can find it at www.taxreformpanel.gov.

CLOUDY AND IMPERFECT

So here's our question for you. Given that the current system is cloudy and far from perfect, what should we do? If we rule out tax hikes, while we try to come up with a fairer, simpler system, we'll lose time and might end up having to make much deeper cuts in spending to get the budget under control. And if we decide to raise taxes, we'll need to be very careful and thoughtful about how to do it.

---

[13] "Press Gaggle by Scott McClellan, Aboard Air Force One En Route Albuquerque, New Mexico," August 11, 2004, available at www.white house.gov/news/releases/2004/08/20040811-4.html.

Lost at Sea: A Short History of Taxing Yachts

It seemed like such a good idea at the time. It's 1990, and the federal government is running in the red. President Bush (the first one) and Congress need to raise money. Better to tax the wealthy than the millions of voters in the middle class. Rich people buy yachts. Lo and behold, Congress passes a 10 percent luxury tax on yachts selling for more than $100,000. The luxury tax also applied to furs, jewelry, and watches costing over $10,000, planes costing over $250,000, and automobiles costing over $30,000. Just a reminder—back in the early 1990s, a $30,000 car was still pretty swish.

Putting a special tax on yachts seemed like a good idea until yacht builders in New England began to go out of business. *Credit: IStockphoto.com*

But while the plan sounded good to a lot of people in Washington and elsewhere, it went badly wrong. The yacht tax was intended to help the federal government out of a funding crunch, but it ended up devastating the boatbuilding industry. Socked by the double whammy of a recession and a hefty tax on their product, boatbuilders saw sales of bigger yachts drop by 70 percent.[14]

---

[14] Shepard W. McKenney, "Taxing Jobs Away," *Washington Post*, November 24, 1992. McKenny was chairman of a yacht-building company in Southwest Harbor, Maine.

Several prestige yacht manufacturers even filed for bankruptcy during the period.[15]

For conservative columnist George Will, the upshot was obvious: "People bought yachts overseas. Who would have thought it?" he wrote in one of his columns.[16] But even those not ideologically opposed to taxes on wealthy Americans had to admit that this one was backfiring.

One boat builder chronicled the cascading impact: "The truth is that while yachts are a luxury for the rich, they are a necessity for American yacht workers. Yachts are in fact great redistributors of wealth. A typical $1 million yacht requires 12,000 labor hours (eight worker years) to build, not counting all the manufactured parts supplied by other domestic industries which provide their own employment, or the considerable labor required to maintain such a yacht. When the buyer pays for the yacht, the money goes to the workers."[17] Not too long after, Republican President Bush joined Democratic Senate majority leader George Mitchell in calling for repeal. In 1993, Congress complied.

This little tale of unintended consequences demonstrates the pitfalls of jumping on easy answers to solve the country's budget problems. It also illustrates some of the subtleties lawmakers (and voters) need to keep in mind when considering raising taxes. Taxes on "luxuries" such as yachts and furs affect more than the buyers. Taxes on "sins" like buying alcohol and cigarettes have advantages of reducing their use (good for people's health), but they, too, affect workers and shareholders in the companies that produce them (not so good for tobacco-raising North Carolina, for example). The effect of raising income taxes is more spread out, but this is definitely not a popular idea. What's

[15] Nick Ravo, "Big Boats Take It on the Chin," *New York Times*, April 14, 1991.

[16] George Will, "Tax Breaks for the Yachting Class," *Washington Post*, October 28, 1999.

[17] McKenney, "Taxing Jobs Away."

more, most economists point out that pushing income taxes up too high leaves people with less to spend on products and services, which also jeopardizes the economy. Broad-based (meaning nearly everyone pays them) "consumption" taxes (essentially sales taxes) are the least likely to harm the economy, according to many economists, but sales taxes are some of the most unpopular taxes of all.

Then, of course, you may decide you'd like to tackle the country's budget problems without raising taxes at all—well, let's just say that there's a lot of slashing to be done. Think you can cut the fat without hitting any bone? You can have at it in Chapter 16, but it's just so much easier said than done.

THE AWOLS (OR, SOME IMPORTANT THINGS WE LEFT OUT OF THIS CHAPTER)

You know those "Great Moments in Music" recordings that serve up snatches of Beethoven and Brahms (rumor has it that there's a "Great Square Inches of Art" parody showing Mona Lisa's smile, but no Mona Lisa). Our Taxes 101 intro left out some important stuff, too. Now that we've covered the basics, there are some other considerations you need to think about. We take up some of these later, but you may want to rev up your search engines and delve into these in more detail yourself.

The "alternative minimum tax." This is the tax nearly everyone wants to cut—liberals, conservatives, and even a fair number

of budget hawks—and there are some pretty good reasons to change it. It's a weird, complicated hyperspace of the tax system originally aimed at the wealthy. The problem is it's now beginning to affect lots of people who are a long way from rich. But fixing the AMT, as it is affectionately called, will mean a big, big loss to the Treasury. We cover the issue in chapter 14.

The estate tax. There's a lot of agreement that Congress should fix the AMT so it only applies to people who are really wealthy, but bring up the estate tax, and you've got a fight on your hands. It's been cut over the last several years, but now there's a big debate on whether to eliminate it or take it back to earlier levels. We discuss the estate tax in chapter 14, but you can also get a good sense of the pros and cons by visiting www.clubforgrowth.org, an organization that urges repeal, and www.responsiblewealth.org, an organization that opposes repeal.

More on fairness and simplification. We just scratched the surface on the fairness and simplification issues here. There's talk, for example, of replacing the income tax with a national sales tax (check out www.FairTax.org for some ideas on this) or a value-added tax (see chapter 9). Former Republican presidential candidate Steve Forbes has long been an advocate of a "flat tax," and he's written about his ideas extensively (www.pbs.org/newshour/bb/congress/forbes_flat_tax.html). Oregon senator Ron Wyden also has a proposal that would reduce the tax code to three tax brackets and allow just a few deductions (yes, the home mortgage interest deduction is one of them).[18] You can visit his Web site (www.wyden.senate.gov) to find out more about his approach.

---

[18] Floyd Norris, "Tax Plans of Candidates Are a Mystery," *New York Times*, June 8, 2007.

Showdown at Tax Gap

Every year, lots of Americans don't pay the government everything they owe in taxes. (No, we didn't mean you. And don't look at us, either.) The IRS estimates the "tax gap," the difference between what people should pay and what they actually pay, at between $312 billion and $353 billion. Between late payments and the IRS actually chasing down tax cheats, the agency says it manages to recover about $55 billion of that, leaving at least $250 billion still out there.[19]

You'll hear a lot about this in the coming years. By Washington standards, this is a "fun fact," because it opens the possibility that the government could go a long way toward closing its annual deficit just by chasing down what's already owed. Which would also mean there's no need for a tax increase or program cuts.

So where is this money, anyway? And why doesn't the government get it already?

One key thing to understand about the U.S. tax system is that it runs on "voluntary compliance," the idea that citizens pay their taxes out of their own free will rather than because they're afraid of the IRS. (No, really. Stop laughing.) Think about it. When you figure out your Form 1040 every year, the government is pretty much relying on you to put down honest numbers. And to look at the tax gap another way, Americans pay about 84 percent of the taxes they owe—a pretty high percentage.

Of course, there are a lot of reasons to fill out that tax return accurately. One is that with many common sources of income, the government has adopted Ronald Reagan's old maxim of "trust, but verify." If you get paid a regular salary or wages, the government withholds the taxes before you even see your check and

---

[19] "Understanding the Tax Gap," Internal Revenue Service fact sheet, March 2005 (www.irs.gov/newsroom/article/0,,id=137246,00.html).

then requires your employer to report the totals (the near-universal Form W-2). More recently, the government started requiring banks and investment firms to report how much you're earning in interest and dividends. IRS studies show there isn't much cheating in these "third-party reporting" areas, because there are too many people looking over your shoulder.

So where does the gap come from? Underreporting the income that nobody's watching (mostly income from partnerships or other small-business activity) accounts for 80 percent of the gap, according to the IRS.[20] In those cases, the IRS has one other tool to chase down the money: the dreaded tax audit.

The problem is that there are fewer audits than there used to be. And, many experts argue, the main reason for that is that Congress told the IRS to be nicer to people. (Again, stop laughing). In the early 1990s, there were congressional hearings into IRS horror stories—property and businesses confiscated for incorrect back-tax claims, people facing years in court to clear up bookkeeping errors, even people driven to suicide. In 1996, Congress passed the Taxpayer Bill of Rights, leading to a major push to help taxpayers avoid mistakes in the first place instead of chasing them down afterward. As a result, audit rates plummeted. In 1984, 1.19 percent of individual returns went through a "face-to-face" audit with an IRS examiner; by 2004, that number dropped to just 0.15 percent. The IRS has been increasing enforcement efforts, but they're still well below the levels of the 1990s.[21]

And most definitions of the tax gap don't even cover big corporations, which have huge accounting staffs and legal teams

---

[20] "New IRS Study Provides Preliminary Tax Gap Estimate," Internal Revenue Service, March 29, 2005 (www.irs.gov/newsroom/article/0,,id=137247,00.html).

[21] Julie Kosterlitz, "Lure of the Tax Gap," *National Journal*, November 4, 2006.

devoted to paying the least possible amount of taxes, within the law. IRS staff time devoted to corporate audits has also been declining, according to an independent study—and auditing big corporations is a time-consuming business.[22] The IRS also says more Americans are using abusive offshore tax shelters to hide their income, but those shelters are also difficult to chase down.

So closing the tax gap isn't exactly found money. It's a lot of hard work and it leads to a fundamental question: How in-your-face do we want the IRS to be? We could audit more people, but who gets audited, like who pays taxes in the first place, has all kinds of political ramifications. We could also expand those third-party reporting requirements to cover more small businesses, but that's going to be a burden to mom-and-pop operations.

Everyone wants to close the tax gap. Everyone thinks cheaters should have to fork over the money they owe. But would you be as enthusiastic about closing the gap if it meant your own chances of being audited went up? (Remember, most audits occur because the IRS suspects something's wrong, but some are triggered by honest mistakes or just as random spot-checks.) So, as we said up top, you can talk tax gap all you want, but closing it is another story entirely. Or as the Rolling Stones so nicely put it, "You can't always get what you want."

---

[22] "Easier Times for Biggest Corporations," Transactional Records Access Clearinghouse, accessed January 14, 2007 (www.trac.syr.edu/tracirs/latest/174/).

CHAPTER 5

And Money Goes...

No government ever voluntarily reduces itself in size. Government programs, once launched, never disappear. Actually, a government bureau is the nearest thing to eternal life we'll ever see on this earth!

—*President Ronald Reagan, TV address, October 27, 1964*

The federal government is vast. Mind-bogglingly vast. With 2.7 million civilian employees and another 1.4 million active-duty military personnel, it dwarfs anything in the private sector.

By contrast, Wal-Mart, the world's largest private employer, has about 1.8 million employees. If you consider the fact that Wal-Mart has yet to field an armored division, they're not too far behind. (But if you think they're tough on competitors now, consider what they could do with cruise missiles.) Yet Wal-Mart is involved in only one business: retail stores. The federal government has millions of tasks it has to perform every day, in a wide range of fields. And the roster isn't limited to just what federal employees do directly. Even more jobs are handled by federal contractors, or conducted by state or local governments with federal

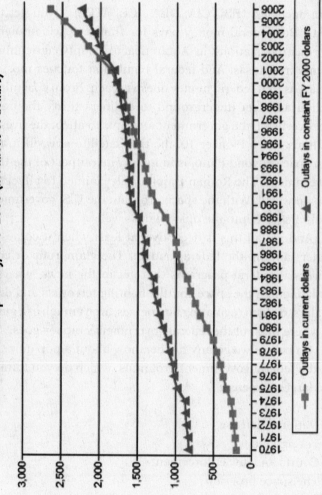

U.S. Government Spending, 1970–2006 (in billions of dollars)

Legend: Outlays in current dollars — Outlays in constant FY 2000 dollars

If you think the federal government is spending more than it used to, you're right, even when using "constant dollars" to adjust for inflation. *Source: Budget of the United States Government, FY 2008* money. Wal-Mart just *seems* like it's everywhere; the federal government really *is* everywhere.

Sure, the military is an obvious federal function, as is the array of law-enforcement and national security acronym agencies: (FBI, CIA, DEA, ICE, ATF . . . you get the idea). But federal money pays for things as wide-ranging as research stations in Antarctica to county agricultural agents in Kansas. And federal regulation touches most of the areas that federal money doesn't reach. From coal mines a thousand feet underground to airliners thirty thousand feet overhead, the government attempts to affect the lives of millions. Even Pioneer 10, the tiny satellite now off in the vast space beyond Pluto, is in its way an outpost of the U.S. government. The Roman Empire only claimed to affect the known world. With the space program the U.S. government sticks its nose into the unknown, as well.

And yet all this is deceptive, at least when it comes to understanding the federal budget. The glamorous or controversial federal programs that get in the news, whether it's launching the space shuttle, fighting terrorists and drug dealers, or even regulating businesses, aren't much of a guide to where most of the federal government's money goes.

So before we go any further, how about a pop quiz? Of all the federal government programs, which do you think is the largest expense:

★ National defense
★ Foreign aid
★ Courts and law enforcement
★ The space program
★ Social Security

Yes, it's a trick question (you probably figured that out already). And to really understand the answer, a little tour is in order, a trip through the federal budget and how much is spent on what. If we're going to start cutting and rearranging, we all need to know where the money is. You probably opened this book with definite ideas about what federal programs could get chucked. You may close the book with those exact same ideas, and that's fine. But you have to know what those cuts will actually do for the budget.

Because for all its complexity, for all its reach, the fact remains that if you gauge it by the federal budget, the main function of the world's greatest superpower is . . .

. . . writing checks to retired people.

Yeah, we know. Surprised us, too. You'd better sit down and have a look at this pie.

Federal Spending, 2006 (in billions of dollars)

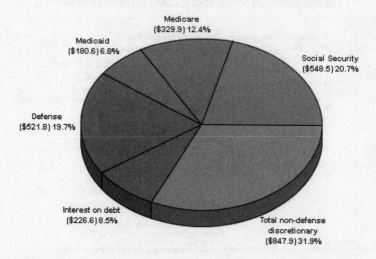

Medicare ($329.9) 12.4%

Medicaid ($180.6) 6.8%

Social Security ($548.5) 20.7%

Defense ($521.8) 19.7%

Interest on debt ($226.6) 8.5%

Total non-defense discretionary ($847.9) 31.9%

Not what you expected, is it? The U.S. government spent more than $2.6 trillion in 2006, and the biggest individual slices were Social Security, defense, and Medicare. *Source: Budget of the United States Government, FY 2008*

Not what you thought, was it? The fact is that the federal government spends about 68 percent of its money on just five things: Social Security, national defense, Medicare, Medicaid, and interest on the money we've already borrowed, thanks to previous deficits. The rest of the budget, from veterans' hospitals to welfare, from small-business loans to office chairs, takes up about a third of the budget.

It gets better (or worse, depending on your point of view). Not all federal government programs are created equal. Some of them are "discretionary," which basically means that what Congress giveth, Congress can also take away. A budget item may be truly vital, like national defense or law enforcement, and still be discretionary. Congress can spend as much or little as it thinks fit.

Federal Spending by Category, 2006 (in billions of dollars)

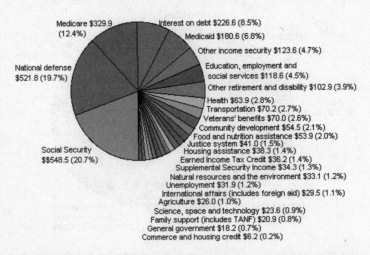

Medicare $329.9 (12.4%)
Interest on debt $226.6 (8.5%)
Medicaid $180.6 (6.8%)
Other income security $123.6 (4.7%)
Education, employment and social services $118.6 (4.5%)
Other retirement and disability $102.9 (3.9%)
National defense $521.8 (19.7%)
Health $63.9 (2.8%)
Transportation $70.2 (2.7%)
Veterans' benefits $70.0 (2.6%)
Community development $54.5 (2.1%)
Food and nutrition assistance $53.9 (2.0%)
Justice system $41.0 (1.5%)
Housing assistance $38.3 (1.4%)
Social Security $$548.5 (20.7%)
Earned Income Tax Credit $36.2 (1.4%)
Supplemental Security Income $34.3 (1.3%)
Natural resources and the environment $33.1 (1.2%)
Unemployment $31.9 (1.2%)
International affairs (includes foreign aid) $29.5 (1.1%)
Agriculture $26.0 (1.0%)
Science, space and technology $23.6 (0.9%)
Family support (includes TANF) $20.9 (0.8%)
General government $18.2 (0.7%)
Commerce and housing credit $6.2 (0.2%)

So what's going on in that budget slice called "discretionary spending"? Only nearly everything people usually think of when we think about the government, including education, veterans' programs, the space program, and a host of other programs. *Source: Budget of the United States Government, FY 2008*

Other programs are "entitlements," services set up so that as long as you qualify, the government has to pay you the money. Social Security and Medicare are both entitlements. So long as you're old enough, you get payments, no matter what.

Sometimes people say you can't cut entitlements. That's not really true. Congress could change the eligibility requirements or trim the payments if it wanted to.

What is true about entitlements is that they're on autopilot. Unlike other programs, Congress doesn't have to review them as part of the budget process every year and make a specific decision to spend more or less on them. What is also true is that Congress rarely has the nerve to change them. There are a lot of people depending on entitlements, and nobody wants to hurt old people, or even suggest a change that might remotely look like they want to hurt old people. Besides, since the entitlements are on autopilot, Congress doesn't *need* to make any decisions. The spending formulas are set, the taxes are collected, the payments are made. So why go looking for trouble?

And, as you've probably already noticed, of the "big-five" budget items, two of them are entitlements. Another slice of the pie, interest on the national debt, isn't an entitlement but is also off limits. If the government doesn't pay the banks, they won't lend the government any more money.

Let's go through the major sections of the budget, slice by slice, and see where we stand.[1]

---

[1] These are all 2006 figures, the most recent available. If you want to get into this in more detail, all the numbers are at www.gpo.gov/usbudget.

Social Security and Medicare Take a Larger Share

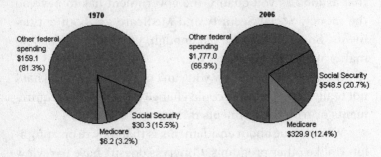

This gives you a real sense of what's going on—Social Security and Medicare are taking up a much larger share of the federal budget than they did a generation ago, and it's only going to continue. *Source: Budget of the United States Government, FY 2008*

Social Security: 20.7 percent of the budget, $548.5 billion

It's no wonder this is the biggest federal program. Nearly 49 million Americans were getting Social Security benefits in 2006. And working Americans who aren't getting Social Security now are paying taxes to support those who are.

That's a key fact to understand about Social Security. It's set up as a "pay-as-you-go" program. People in the workforce now pay taxes to cover the check your grandma gets every month. And the money Grandma paid in? That went to pay for people who were retired while she was working. Social Security isn't a savings or investment program. There's no account with your name on it where your Social Security taxes are actually set aside. You do, of course, get a "statement" tallying up how much you've paid into the system, but almost all the money has already been used to pay benefits for older people.

But what about this Social Security trust fund politicians talk about? Currently, Social Security runs a surplus—the taxes we all pay in are more than enough to cover all the people currently getting checks, and there's some money left over. So the government is supposed to bank that extra money in a trust fund. Ideally, that fund is a hedge against the long-term problem facing Social Security, which is that once the baby boomers start retiring there will be a lot more people drawing checks than paying in taxes. At some point, however, the trust fund will run dry (currently projected to occur in 2042). Lots of people think Social Security will be bankrupt at this point, but that isn't really accurate. The system will still operate, but there will only be enough money coming in to pay for about 75 percent of the benefits needed. Which is not, technically speaking, bankruptcy, but it isn't terribly comforting, either. Just think what would happen if all the older people on Social Security now saw their benefits drop by 25 percent.

There's only one problem with the Social Security trust fund. It mainly exists on paper. Rather than let the surplus Social Security money sit in the bank, the government has been borrowing it for day-to-day operations. The government promises to pay the fund back, and there's no reason to doubt it, given how unpopular it would be to fail to pay benefits. But that still means the government will have to come up with the money at some point.

National defense: 19.7 percent of the budget, $521.8 billion

This pretty much means what you'd think it means: the Pentagon and the nation's far-flung national security apparatus, including the intelligence agencies. It also means the wars in Iraq and Afghanistan, although you won't find a line item marked "Iraq" in there. So how can you be fighting a war and not have it reflected in the defense budget? Up until 2006, the Bush administration relied on "supplemental appropriations" or special spending bills to cover the cost, which the Congressional Budget Office estimates at $602 billion for "the global war on terror" through mid-2007.[2] Trying to figure out exactly how much has been spent is difficult, because the money is spread out among different accounts. (See "The Fortunes of War: Why bringing the Troops Home Won't Balance the Budget,"on page 94).

With the 2007 budget, the Bush administration started including war expenses in the president's budget request, asking for $93.4 billion in 2007 and anticipating another $141.7 billion in 2008.

Medicare: 12.4 percent of the budget, $ 329.9 billion

Medicare is the government's health insurance program for the elderly. Like Social Security, it's an entitlement. If you're over the age of sixty-five, you can get it, and almost everyone does—some 43 million Americans. Essentially, Medicare has all of the challenges of Social Security, plus some twists of its own.

Medicare has multiple "parts." Part A, financed by a payroll tax paid equally by employees and their employers, is a hospital insurance program that covers most inpatient hospital costs. Older Americans can elect to enroll in Part B supplementary insurance, which covers physician and outpatient services. Recipients pay premiums for this, and the rest

---

[2] Robert A. Sunshine, "Testimony on Estimated Costs of U.S. Operations in Iraq and Afghanistan and of Other Activities Related to the War on Terrorism," Congressional Budget Office, July 31, 2007 (www.cbo.gov/ftpdocs/84xx/doc8497/07-30-WarCosts_Testimony.pdf).

comes from the government. Medicare spending that isn't covered by payroll taxes and premiums comes directly from the federal government, up to 45 percent of the total program bill. There's also the newer Medicare Part D, passed in 2003, which helps older Americans buy prescription drugs.[3]

Medicare has two unique problems of its own. One is health care costs. Just as all other health insurance is becoming more expensive, so is Medicare. Scientists keep finding new treatments for older people, who are living longer and longer, which is wonderful—and expensive. And it would be hard to find someone on the political left or the political right who doesn't think that there is some significant amount of waste and duplication in the health care system (they, of course, have very different explanations and cures for it, but more about that in chapter 9). The other is that Medicare's trust fund is being depleted even faster than Social Security. So revenues will fall short of expenses in 2019, much sooner than Social Security. The Part D Medicare prescription drug program, while well-intentioned, will add another $578 billion in expenses to the program from 2007 to 2013.[4] The addition also brought the day of reckoning seven years closer.

Medicaid: 6.8 percent of the budget, $180.6 billion

The health insurance program for the poor is actually a partnership between the federal government and the states.

---

[3] Wondering where Part C is? Yes, there is one. It's the Medicare Advantage plan, in which seniors can get all their services from a single HMO/PPO.

[4] Congressional Budget Office, "The Budget and Economic Outlook: Fiscal Years 2008 to 2017," January 2007, available at www.cbo.gov/ftpdocs/77xx/doc7731/01-24-BudgetOutlook.pdf.

The states run it, and the federal government pays part of the costs. So while people under a certain income level are eligible for it, the program isn't running on autopilot like Social Security and Medicare. The federal government can cut how much it pays if it wants. Of course, that sticks the states with the bill. Your federal taxes might go down, but your state taxes might go up. What you make up on the Ferris wheel you lose on the merry-go-round.

Interest on the debt: 8.5 percent of the budget, $226.6 billion

This is pretty much the same as the minimum payment on your credit card. This is how much the government has to pay to banks and holders of Treasury bonds for the money we've already borrowed—and to keep the door open to borrow more.

But what about everything else? What about all that foreign aid, the space shuttle, welfare? Have another look at the pie. See all those very narrow slices? That's where all that spending is, from national forests to disaster relief, from weather satellites to interstate highways. It just goes to show that the most obvious government functions aren't necessarily the biggest ones.

Just because these programs are small proportions of the federal budget, doesn't mean they ought to be kept going. Government waste is real enough. Some programs don't work, some programs should be lower priorities than they are. But you can't get very far in tackling this problem without knowing where the money is.

Prime Cuts or Deli Slices?
What It Takes to Make a Dent in the Deficit

Some government programs, like some people, are just inherently annoying. Nearly everyone has some government function that they think is a waste of money or just plain wrong. If all those programs were cut, surely the deficit would go away. Right?

Well, let's test it. When we do focus groups on the deficit, people are quick to come up with suggestions of things to cut. Here are a few of the ones we hear regularly. Let's walk through them. Check off any you think should go and we'll total

Astronaut Pete Conrad unfurls U.S. flag on the moon. Spending on science, space, and technology takes up about 1 percent of the budget. *Credit: NASA Photo*

them up at the end. And to make the cuts as deep as possible, we're taking on whole slices of the budget pie, eliminating entire categories of spending. In other words, we've put the baby in with the bathwater. If you cut a category, you're cutting all of it.

SCIENCE, SPACE, AND TECHNOLOGY

We rarely seem to get any Trekkies in our focus groups, because the space program is usually one of the first to be jettisoned. The basic argument is "Why are we shooting rockets into space when there are so many problems to be dealt with here?" Defenders of government research funding, particularly "basic research" that doesn't have a short-term payoff, say it expands human knowledge and pays off in the long run. For our purposes here, we're throwing in the entire science budget, including the National Science Foundation, which ranges from physics to social sciences to biology. Some medical research funding lies elsewhere in the budget.

Total savings: $23.6 billion, less than 1 percent of the budget

THE ARTS AND HUMANITIES

The beauty of the National Endowment for the Arts is that if it ever wants to get into trouble, it has a simple way of doing it: just fund something avant-garde. During the 1990s, the agency's support for Robert Mapplethorpe and other push-the-envelope artists made it a favorite target of conservatives. The National Endowment for the Humanities has been less controversial over the years. Supporters of these programs say the arts enrich our lives and even provide economic benefits to communities. But many argue that the government shouldn't be in the business of funding museums, symphonies, artists, writers, linguists, and historians. Artists should get their money from the private or nonprofit sectors.

Total savings: NEA, $124 million, NEH, $141 million, together 0.009 percent of the budget

FOREIGN AID AND INTERNATIONAL RELATIONS

Average Americans rarely have much good to say about foreign aid. After all, we have lots of problems at home. And there are many critics who say aid programs are poorly run and often wasted in corrupt countries. There are a lot of people we could quote on this, but we've always been partial to the cartoon show *Pinky and the Brain,* about a superintelligent lab mouse bent on world domination. In one episode, the Brain cons the U.S. government out of billions in foreign aid by pretending to lead the strategic country of "Brainania." At the check ceremony, the Brain announces: "Mr. President, I thank you for your friendship and for this $19 billion aid check. The friendship I will cherish; the money I will spend on polo ponies and cruise missiles."

Advocates of foreign aid don't have a cartoon for their point of view, but they do point out that the United Sates gives less proportionately than many other wealthy countries, while the poor in places like Africa and Asia are far poorer than even the most needy American (fully 1 billion people in the world live on $2 a day or less). Plus, foreign aid buys us goodwill, which is a useful asset in diplomacy and often in short supply. For our purposes, we're being hard-core and cutting all international relations funding—not just the aid, but shutting down all the embassies and the State Department itself.

Total savings: $29.5 billion, 1.1 percent of the budget

THE PROGRAM FORMERLY KNOWN AS WELFARE

When the federal government overhauled welfare in 1996, few people were sorry to see it go. Surveys proved that even people on welfare didn't like the way the program ran. And the idea of "welfare dependence"—that living on government assistance was being passed from parents to children—scared and offended most Americans. Since the changes, welfare rolls have been cut dramatically and there is now a five-year lifetime limit on benefits. Even though federal food-assistance programs like food stamps and free school lunches weren't affected, enrollment in those dropped, too. Most say the system now runs much better, but no one would call it popular. If you're being hard-nosed about cutting the deficit, this is always an item.

Total savings: Temporary Assistance for Needy Families; food and nutrition assistance, including food stamps: $70.9 billion, 2.67 percent

And it all adds up to? A grand total of 4.08 percent of the budget.

Not a big dent, is it? Granted, this isn't solely about saving money. It's perfectly all right to get rid of a program just because most Americans think it's a waste of time or that the money can be put to better use. And in other cases, a program should be kept because it's important, even if it's costly. But we're worried about balancing the budget here, and while every little bit helps, cuts like these aren't going to do the job. It's just not going to be this easy.

The Fortunes of War: Why Bringing the Troops Home Won't Balance the Budget

When we conduct focus groups on government finances, one of the first things we often hear people say is "If we weren't in Iraq, we wouldn't have a deficit." Or, depending on their viewpoint, "We have to be in Iraq and that costs money, so we have to put up with the deficit." In other words, wars are expensive, emergencies are emergencies, and this is no time to worry about the deficit.

Wars certainly are expensive, which is why nations have been known to go broke fighting them. At the moment, however, that's not what's happening to the United States.

As of July 2007, the Congressional Budget Office estimates Congress has authorized $602 billion for military and diplomatic operations in Afghanistan, Iraq, and the rest of the global war on terrorism since 2001. At least 70 percent of the total has been allocated to Iraq, according to the CBO.[5] Because of the way the government breaks out its billing, it isn't clear how much of that

---

[5] Sunshine, "Testimony on Estimated Costs of U.S. Operations in Iraq and Afghanistan and of Other Activities Related to the War on Terrorism."

the Pentagon has actually spent, but government officials estimate the monthly "burn rate" at between $9 and $10 billion.[6]

And for the future? President Bush's 2007 budget request asked for $93.4 billion for military spending for Iraq and counterterrorism in 2007 and projects another $141.7 billion for 2008. As of August 2007, the CBO expected a $158 billion deficit in 2007.[7] So even if we lived in a world where we could withdraw from Iraq overnight, with no winding-down period, where just red-penciling the Iraq war out of the budget would make it go away, it still wouldn't close the deficit. And it would do absolutely nothing about the long-term problems caused by the aging of the boomers.

Frankly, the most significant fiscal fact about Iraq in President Bush's budget request was that the war was included at all. Up until the 2007 budget request, *none* of the president's budget requests had included funding for Afghanistan or Iraq, for five straight years. So you may be asking, what kind of nutball accounting system makes that possible? How could the government be fighting a war costing hundreds of billions of dollars and not have that clearly reflected in the budget?

Welcome to the world of supplemental or "emergency" appropriations, which in reality are not completely insane. Yes, the government has a budget, but things come up suddenly, like wars and hurricanes, and when they do Congress can pass special supplemental appropriations to pay for them. The formal budget takes a year to prepare and months to get through Congress and wars move faster than that, as former defense secretary Donald Rumsfeld said. "Supplemental appropriations are put together much closer to the time the funds will actually be used," Secretary

---

[6] Ibid.

[7] Congressional Budget Office, "The Budget and Economic Outlook: An Update," August 2007, available at www.sco.gov/ftpdocs/85xx/doc8565/08-23-Update07.pdf.

Rumsfeld told Congress. "This allows a considerably more accurate estimate of costs, and, importantly, much quicker access to the funds when they are needed, without having to go through . . . contortions where we are forced to rob other accounts and distort good business practices."[8]

Fair enough, so far as it goes. But that also means that it's been very difficult to know how much the wars will cost, or how much has actually been spent. Don't take our word for it—the Iraq Study Group Report reached the same conclusion. "Detailed analyses by budget experts are needed to answer what should be a simple question: 'How much money is the President requesting for the war in Iraq?'" the report said.[9] Even during Vietnam, when the Johnson administration was trying hard not to let on how big or how long the war would be, war costs were included in the regular budget. When the administration guessed wrong, it asked for supplemental funds.[10]

So President Bush's decision to put war costs in the regular budget request is a major step forward for good fiscal management, even if the fortunes of war mean the final numbers end up quite different. But that doesn't affect the fundamental fact that bringing the troops home, by itself, isn't going to solve the financial problem. Peace is certainly cheaper than war, but not cheap enough to make the books balance—at least not in this case.

---

[8] Donald Rumsfeld, Secretary of Defense, FY 2006 Supplemental Request Statement Before the Senate Appropriations Committee, March 9, 2006 (www.defenselink.mil/speeches/2006/sp20060309-12630.html).

[9] Iraq Study Group Report, Recommendation 72, December 6, 2006 (www .usip.org/isg/iraq_study_group_report/report/1206/index.html).

[10] "Military Operations: Precedents for Funding Contingency Operations in Regular or in Supplemental Spending Bills," Congressional Research Service, June 13, 2006.

CHAPTER 6

Social Security and Medicare– and Why Closing the Deficit Isn't Enough

The Social Security trust fund is what I call a fiscal oxymoron.
It shouldn't be trusted, and it's not funded.

—*Pete Peterson, former U.S. secretary of commerce, 2005*

We've said it before, and we'll say it again: Social Security and Medicare are pay-as you-go programs. Payroll taxes collected from those of us who work cover retirement and health care for those of us who are old. That's how these systems were designed decades ago. Unfortunately, an arrangement that has worked very smoothly for quite a while is about to encounter some mammoth bumps in the road. Here's why.

No. 1: The boomers are about to go on Social Security. The baby-boom generation is a lot larger than generations after it (even with the twin beds and all, couples managed to make a lot of babies back in the 1950s), so fairly soon there will be a lot more people collecting Social Security and a lot fewer paying into it. You can see how this could cause a problem.

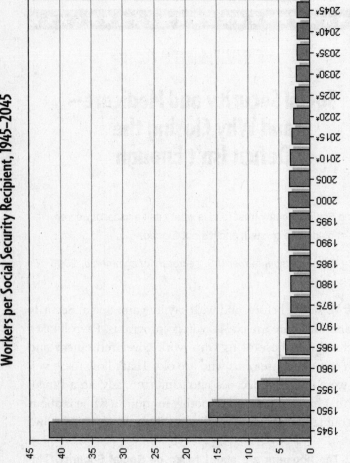

Workers per Social Security Recipient, 1945–2045

Social Security is designed as a "pay-as-you-go" system. But the problem is that if not enough people are paying in, the system won't go. And the proportion of workers to each Social Security recipient is falling. *Source: Social Security and Medicare Trustees Report, 2006*

No. 2: People are living longer. When Social Security began in 1935, there were only about 7.5 million Americans over age 65. Today, there are about 36 million Americans over

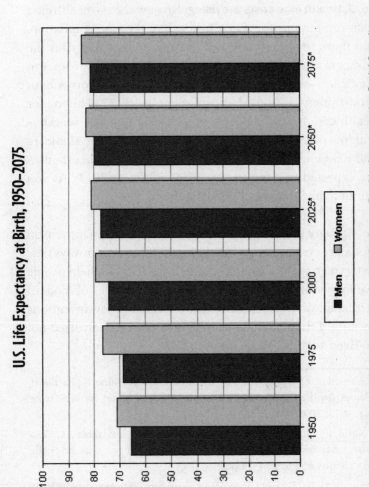

U.S. Life Expectancy at Birth, 1950–2075

Don't get us wrong, this is good news: people are living longer. But this does mean that Social Security and Medicare costs are going to keep rising.   *Source: National Center for Health Statistics, CDC*

65.[1] Or to put it another way, the average life expectancy in 1940 was a little under 64 years. A child born in 2004 can expect to live 77.8 years.[2] This is not brain surgery. When more people live longer, collect Social Security, and get health care paid for by Medicare, it costs more money.

No. 3. Health care costs are rising. No news here—health care costs are famous for rising faster than the rate of inflation, and that's for everybody, not just the government. But the numbers are beginning to be a little scary when it comes to Medicare—scary enough to alarm experts who worry about these things. In 2006, Medicare spent nearly $330 billion for health care for older people (yes, just in that one year, lots and lots of money). With more people eligible for Medicare and rising health care costs, the program's annual expenses are expected to jump to over $700 billion by 2017.[3] As you can imagine, this is not good for the budget.

No. 4: Every year there's more health care to buy. One reason Medicare costs go up so quickly is that there are always new tests, treatments, machines, and drugs that can help people live longer and more comfortably—the miracles of modern medicine. We expect Medicare to cover these new inventions because if they can help people, it would be uncivilized not to. Even so, this costs a lot of money.

---

[1] Statement of Social Security commissioner Jo Anne B. Barnhart, released by the Social Security Administration, States News Service, August 11, 2005.

[2] National Vital Statistics Reports, vol. 54, no. 14, Table 11, "Life Expectancy by Age, Race, and Sex," available at www.cdc.gov/nchs/data/nvsr/nvsr54/nvsr54_14.pdf.

[3] Congressional Budget Office, "Fact Sheet for CBO's March 2007 Baseline: Medicare," available at www.cbo.gov/budget/factsheets/2007b/medicare.pdf.

No. 5: We've expanded benefits. Over the years, the country has added improvements to both programs (see "Social Security and Medicare: A Quick History," page 109). The most recent example is the prescription drug coverage Congress added to Medicare in 2003. Since a lot of elderly Americans were having trouble paying for drugs, Congress passed legislation to help them—legislation supported by Democrats, Republicans, and nearly nine in ten members of the public.[4] The drug plan, which is available to every Medicare recipient regardless of income, is expected to add $518 billion to the cost of Medicare between 2007 and 2013.[5] Providing drug coverage for all Medicare recipients—not just those with lower incomes—was also supported by most Americans.[6]

Together, these trends are going to put a real squeeze on Social Security, Medicare, and the federal budget. There's really no way around it.

---

[4] According to a Princeton Survey Research Associates/Pew survey, in June 2001, 89 percent of Americans favored making prescription drug benefits part of Medicare. Just 8 percent of Americans opposed it. "The vast majority of Americans say they favor making prescription drug coverage part of Medicare and three-quarters say it should be a top priority." Public Agenda Online (www.publicagenda.org/issues/major_proposals_detail.cfm?issue_type=medicare&list=3), accessed March 4, 2007.

[5] Congressional Budget Office, "The Budget and Economic Outlook: Fiscal Years 2008 to 2017," January 2007, available at www.cbo.gov/ftpdocs/77xx/doc7731/01-24-BudgetOutlook.pdf.

[6] According to a CBS/New York Times poll in June 2001, 62 percent of Americans wanted to make the coverage available to all Medicare recipients versus 35 percent who wanted to provide coverage for only low-income Americans. Cited from Public Agenda Online (www.public agenda.org).

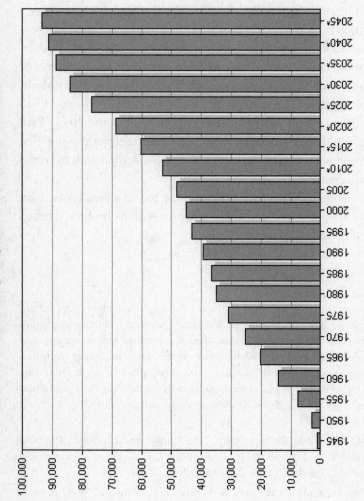

Number of Social Security Recipients, 1945–2045 (in millions)

This is probably no surprise, but the chart brings it home—more and more people are getting Social Security. *Source: Social Security and Medicare Trustees Report, 2006*

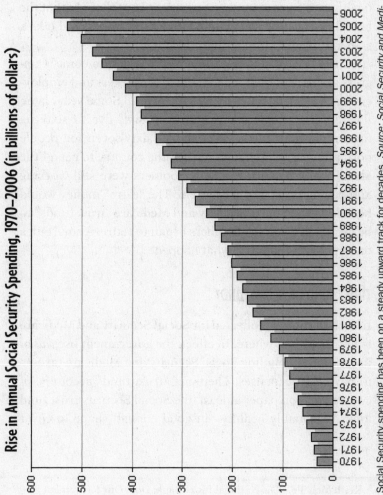

Rise in Annual Social Security Spending, 1970–2006 (in billions of dollars)

Social Security spending has been on a steady upward track for decades. *Source: Social Security and Medicare Trustees Report, 2006*

DIDN'T SOMEONE THINK OF THIS?

Since it's been obvious for years that the baby-boom generation is gigantic, that Americans are living longer, and that
health care costs are rising, you might have thought that the
big "they" would have planned for this. Congress did take a
stab at it back in the 1970s and 1980s.

In 1977, anticipating the money crunch to come, Congress almost doubled the payroll taxes workers and employers pay for Social Security and Medicare.[7] Some years later,
the retirement age was raised from sixty-five to sixty-six
for people born after 1943, and to sixty-seven for people
born after 1960.[8] The idea was for the country to get a little
ahead while most of the baby boomers were still working
and paying taxes into the system. The "extra" money would
be held in the Social Security and Medicare "trust funds" to
be available when the boomers began to retire. A smart idea
on paper, but that's not what happened.

TRUST FUND OR SLUSH FUND?

In fact, the money collected for Social Security and Medicare
is not "held" anywhere. In effect, the government uses *all* of
its income—including Social Security and Medicare taxes—
to pay *all* its expenses. There are "trust fund" accounts, of
course, and, on paper at least, the Social Security trust fund
looks reasonably healthy—in good enough shape to cover

---

[7] Keith Melville, "The National Piggy Bank: Does Our Retirement System Need Fixing?" *National Issues Forums*, 1996, p. 3.

[8] Social Security Online, Retirement Planner, "Retirement Benefits
by the Year of Birth," available at www.socialsecurity.gov/retire2/age
reduction/htm.

projected benefits for seniors until 2041.[9] (We're not talk-ing about Medicare in this section because even on paper it's already in trouble.) But as your mom always told you, appearances can be deceiving. Over the years, the govern-ment has dipped into the Social Security account to cover other kinds of expenses and to avoid having to raise taxes. The government gives the trust fund Treasury bills in return for what it "borrows."

The problem is that Social Security is going to start needing the money it's lent out to other parts of government in about ten years or so,[10] and the U.S. government itself is in the hole nearly every year. It's like you took money from your retirement savings to lend to your good friend Sylves-ter, and he's given you IOUs in return. Sylvester's a decent guy, and his word has always been good, but every year he spends more than he makes, and he's already maxed out his credit cards. Now you need to retire, and those IOUs are just nice little pieces of paper.

REMEMBER THE "LOCKBOX"?

When Social Security starts to redeem its Treasury bills (which it will have to do to pay benefits it has promised to people who paid their taxes into the system for years), the government will have to cut other expenses, raise taxes, or borrow even more money—and probably more and more of it from abroad.

---

[9] Social Security and Medicare Board of Trustees, Status of the Social Security and Medicare Programs, A Summary of the 2007 Annual Reports, "A Message to the Public" (www.ssa.gov/OACT/TRSUM/trsummary.html).

[10] Statement of Secretary Henry M. Paulson Jr. on the 2007 Social Security and Medicare Trust Fund Reports, CQ Federal Department and Agency Documents, April 23, 2007.

Over the years, some experts urged the government to invest the money in the "trust fund" in the stock market, or to allow workers to invest the "extra money" in personal accounts, but these ideas never caught on. When Al Gore was running for president in 2000, the then vice president argued so repeatedly and earnestly for keeping the "extra" money in a "lockbox" (a concept backed by some Republicans as well, including the then Texas governor George Bush)[11] that some comedians began to make fun of him. When the *Boston Herald* organized a focus group of typical voters to watch the vice president debate Governor Bush, the newspaper reported that "Gore appeared to bore viewers with his repetitive comments about a Social Security 'lockbox.'"[12] And given the sheer size of the coming Social Security shortfall and the tendency of politicians to wriggle around arrangements they don't like, some experts said the lockbox idea wouldn't make much difference anyway.[13] Whatever the case, "surplus" Social Security money has routinely been available to cover other government expenses. So when it comes to the Social Security trust fund, there's really not that much "there" there.

THE POLITICAL EQUIVALENT OF CATNIP

For most elected officials—regardless of their political party—being able to "borrow" based on the "surplus" in

[11] CNN.com, "Bush, Gore Continue Heated Debates on Private Social Security Accounts," AllPolitics, May 16, 2000, available at http://archives.cnn.com/2000/ALLPOLITICS/stories/05/16/campaign.wrap/index.html.

[12] Steve Marantz, "Focus group unmoved by debate," *Boston Herald*, October 5, 2000.

[13] See for example, Maya MacGuinas, "Lock Boxes Are Too Easily Unpicked," *Financial Times* (London), August 18, 2000.

the "trust funds" to cover expenses other than Social Security and Medicare is like catnip to a cat. They don't have to ask Americans to pay higher taxes to pay for things like the space program or food stamps or money for special education or agricultural subsidies. Quite the opposite; they can cut taxes and enjoy all the political popularity that brings.

So here we are. Even now, all those boomers are starting to retire in bigger numbers. They (or we, as the case may be) are going to start needing hip surgery and walkers and blood pressure medicine. There will be fewer younger Americans working and paying taxes. So what's a country to do?

Up to now, the country has shown few signs of grappling with the real solutions, which will require some form of trimming benefits and/or raising taxes—or some combination of the two. And if we don't face up to the problem soon, the country is going to get itself into a horrendous financial and political mess.

YEAR 2040: THREE SERIOUSLY BAD CHOICES

So let's say we just sit back and wait for the worst to happen. Here are the three seriously bad choices we'll face.

Choice One

To have enough money to cover the promised Social Security and Medicare benefits for the boomers, we could slash nearly everything else in the budget—college loans, national parks, the Centers for Disease Control, even homeland security and defense. Some experts, including the Government Accountability Office, have calculated that by 2040 or so, if we do nothing, nearly every tax dollar collected will be needed to pay for Social Security, Medicare, and interest on the national debt.

Choice Two

We could place really punishing taxes on working people. These will be people who are trying to raise families, send their kids to college, and pay off the mortgage. And the taxes would have to be very high.

Choice Three

Or we could—suddenly and in a big financial panic—cut the benefits older people have been counting on. These will be people who have themselves paid Social Security and Medicare taxes for decades. Many will be too old to go back to work, even as a greeter at Wal-Mart. Many will be sick and frail.

It's a bleak scenario, but if we can get our collective heads out of the sand, we have the time and the resources to avoid all this. The real question is whether we can stop the sloganeering and electioneering, positioning and gamesmanship, long enough to make some sensible decisions about what to do.

Setting Out the Welcome Mat : Could Immigration Solve the Problem?

Theoretically, one way the United States could produce some extra revenue is by bringing more immigrants into the country to work and pay Social Security and Medicare taxes. We're now living in one of the great eras of immigration, with the Census Bureau estimating that more than one in ten Americans is foreign born.[14] So the country probably wouldn't have any trouble attracting more people if it wanted.

In the mid-1990s, PBS's *Think Tank* host Ben Wattenberg pro-

---

[14] Luke J. Larsen, "The Foreign-Born Population of the United States, 2003," U.S. Census Bureau, August 2004 (www.census.gov/prod/2004 pubs/p20-551.pdf).

posed bringing in an extra 16 million immigrants (about twice the population of New Jersey) to create an "artificial generation of young adults" to help pay the retirement and health care costs of the boomers.[15] Wattenberg's idea is logical enough. Since our problem stems from the fact that the boomer generation is huge while subsequent generations are a little skimpy by comparison, increasing the number of younger workers would make sense.

But would the American public ever support this kind of solution, even if the experts all agreed on the numbers? According to polls, fewer than one in five Americans supports increased immigration.[16] Surveys show that a lot of the public ire about immigration is directed at illegal migrants, rather than lawful immigrants who "play by the rules," but there's not much evidence of a groundswell to throw the doors open even wider.[17]

What many people don't realize is that, financially speaking at least, the Social Security system is already benefiting from illegal immigration. Since you can't get a legitimate job without a Social Security number, lots of illegal immigrants buy fake Social Security cards with made-up or stolen numbers. They cost about $150 on the streets of Los Angeles. Since their Social Security numbers aren't real, they can't actually claim benefits, but the money still gets deducted from their paycheck and used. [18]

---

[15] Ben Wattenberg, "The Demographic Deficit," *Baltimore Sun*, December 15, 1995.

[16] Gallup Poll, June 8–25, 2006, "Thinking now about immigrants—that is, people who come from other countries to live here in the United States: In your view, should immigration be kept at its present level, increased or decreased?" Increase: 17 percent; decrease: 39 percent; present level: 42 percent; unsure: 2 percent.

[17] "Red Flags on Immigration," Public Agenda Online, (www.publicagenda.org/issues/red_flags.cfm?issue_type=immigration) accessed March 4, 2007.

[18] "Illegal Immigrants Are Bolstering Social Security with Billions," *New York Times,* April 5, 2005.

It's not clear exactly how much money this adds up to, but the Social Security Administration says in 2002 there were 9 million W-2 forms filed with incorrect Social Security numbers. Some are no doubt simple mistakes, but Social Security officials estimate three-quarters of them might be illegal immigrants. That accounted for $6 to $7 billion in Social Security revenue and another $1.5 billion for Medicare.

The IRS also believes it's getting a fair bit of revenue from illegal immigrants. The agency offers an Individual Taxpayer Identification Number to allow foreigners who don't have a Social Security number to pay taxes. The IRS issued 1.5 million ITINs in 2006, and agency officials think many of these went to illegal immigrants who believe paying their taxes helps keep them out of trouble in the United States. Between 1996 and 2006, the total tax liability of ITIN holders accounted for about $50 billion.[19]

The debate over what to do about fake Social Security cards and ITINs tells you something about how divisive an "immigration solution" would be. Some support using these filings to track down illegal immigrants and deport them. Those who want a more open immigration policy say it's unfair to make illegal immigrants pay taxes for a system that will never give them benefits. From a purely financial point of view, both arguments have problems: deporting illegal immigrants out of the country would take some ghost Social Security taxpayers out of the system; legalization would allow more beneficiaries in.

What's more, there are massive disputes among experts about the numbers of immigrants and how much money they would actually bring into the system, not to mention the long-term advantages and disadvantages of changing their status. In 2005, the Social Security Advisory Board concluded that increased immigration would be helpful, but "does not view immigration as a panacea—or free lunch— for

---

19 "Illegal Immigrants Filing Taxes More Than Ever," Associated Press, April 13, 2007.

saving Social Security."[20] When it looked at the controversial immigration bill put forward in 2007, the Congressional Budget Office estimated it would "exert a relatively small net effect on the federal budget balance over the next two decades." Yes, more tax money would come into Social Security, but that would be largely offset by increased spending on immigration enforcement and other areas.[21]

Our best advice on this? Regardless of what you think about the larger immigration debate, don't hold your breath waiting for immigration to get us out of our Social Security and Medicare predicament.

---

[20] Social Security Advisory Board, "Issue Brief No. 1: The Impact of Immigration on Social Security and the National Economy," December 2005.

[21] Congressional Budget Office, Cost Estimate, June 4, 2007; Senate Amendment 1150 to S. 1348, the Comprehensive Immigration Reform Act of 2007, as amended by the Senate through May 24, 2007 (www.cbo.gov/ftpdocs/81xx/doc8141/05-23-Immigration.pdf).

SOCIAL SECURITY AND MEDICARE: A QUICK HISTORY[22]

1930   Only 15 percent of workers are covered by retirement plans, many of which collapsed during the Great Depression.

---

[22] Adapted with permission of the National Issues Forums and Public Agenda, "The National Piggybank: Does Our Retirement System Need Fixing," by Keith Melville, 1996.

1935	President Franklin D. Roosevelt signs the Social Security Act at a time when the average life expectancy is sixty-one.
1939	Children of retired workers and surviving children of deceased workers become eligible for benefits.
1954	Social Security is expanded to include agricultural and self-employed workers.
1956	Coverage is extended to disabled workers. Women become eligible for certain benefits at age sixty-two, rather than having to wait until they are sixty-five.
1961	Men become eligible for early retirement benefits at age sixty-two.
1965	President Lyndon B. Johnson signs Medicare into law at a time when only 56 percent of older Americans have health insurance.
1972	Social Security is adjusted, ensuring that benefits rise with the cost of living. Medicare benefits are extended to some people under sixty-five with disabilities.
1977	As costs rise, Congress slows the growth of benefits and nearly doubles the payroll taxes that support these retirement programs.
1983	Congress increases the retirement age in several phases by 2027, from sixty-five to sixty-seven.
1993	Congress increases the Medicare payroll tax by making all earnings taxable rather than just the first $135,000.
2003	President Bush signs a bill adding prescription drug coverage to Medicare.
2007	Medicare Part B, which covers doctors' visits, sets premium "surcharges" for more affluent seniors to help hold down costs.[23]

---

[23] Robert Pear, "Government Sets Higher Medicare Rates and New Surcharge," *New York Times,* September 13, 2006.

The Social Security and Medicare trustees report that Social Security will begin to pay out more than it collects in taxes in 2017.[24]

Medicare's finances are so troubled that the trustees issue a "Medicare funding warning" requiring the president to "propose legislation that responds to this warning" in the 2009 federal budget.[25]

---

[24] Social Security and Medicare Board of Trustees, Status of the Social Security and Medicare Programs, "A Message to the Public," available at www.ssa.gov/OACT/TRSUM/trsummary.html.

[25] Ibid.

Blinding Us with Science: The Debt and GDP

Since the point of this book is to explain the budget problems facing the country as clearly and simply as possible, we don't always use the same terms and concepts as the experts. But there is one concept that we want to draw your attention to because you may come across it when candidates and experts talk about the budget and the debt. It's a very useful concept for economists, but it can be confusing and even a little misleading to the rest of us.

Experts often talk about the deficit and the debt as a percentage of the country's gross domestic product, or GDP. This is the total amount of goods and services produced in the United States—in effect, the total size of the U.S. economy. Most economists say this is the best measure of how the debt affects the entire economy. The bigger the debt is in relation to GDP, the bigger the impact it's likely to have. This is generally true about a lot of things in economics, which is why you'll also see the GDP yardstick used in relation to health care spending, defense spending,

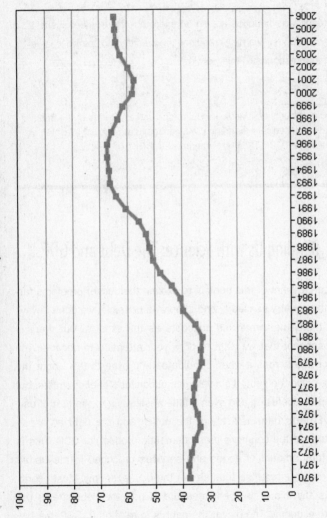

U.S. Debt as Percentage of GDP, 1970–2006

Economists say the country's debt as a percentage of the gross domestic product (GDP) is generally the best indicator of when the national debt gets out of hand. This chart combines both debt held by the public (the Treasury bonds anyone can buy) and intergovernmental debt, such as when the government borrows from the Social Security trust fund. *Source: Budget of the United States Government, FY 2008* taxes paid, and any number of other things. (And not just in the United States, either. It's a standard measure internationally.)

This is just a way of putting big numbers in perspective, in much the same way personal finance experts tell borrowers they shouldn't spend more than one-third of their income on housing. If your mortgage payments are higher than that, they'll eat up money you need for other things. Same thing with the national debt—if it's too big a slice of GDP, other things have to suffer.

In 2006 the "debt held by the public" was about 37 percent of GDP, and the good news is that the country's debt has actually been a bigger share of GDP in the past. Back in 1946, when the country was just starting to pay off all the money it borrowed to fight World War II, debt was at 108 percent of GDP. In 1975 it was down to 25.3 percent. The other piece of good news is that the Congressional Budget Office expects this percentage to go down to 21.2 percent by 2017.

Sounds reassuring, doesn't it? And there are plenty of people who'll tell you, based on these figures, that debt is actually under control, and you don't need to worry about it. Unfortunately, we think those people are wrong, for two reasons.

One is our standard warning about projections. By law, CBO has to make its projections based on current legislation, so it has to assume that all of the Bush tax cuts set to expire at the end of 2010 will actually expire—that Congress won't renew or extend a single one of them. It also has to assume that there won't be any major new spending programs or any additional tax cuts over the next ten years. The poor CBO numbers crunchers must have to put their fingers in their ears when they hear all the candidates running for president and Congress talking about what they are going to do if they win the election.

The second reason is more critical: *"Debt held by the public" isn't all the debt there is.* That doesn't include any of the "intragovernmental" borrowing the government has done from the Social

Security and Medicare trust funds. It only includes the Treasury notes sold to people and institutions here and abroad. There is an actual, legitimate rationale for this. Since the government is borrowing from itself, these loans don't compete in the credit market with your mortgage and car loan. So they're treated differently, just the same way accountants treat internal loans in large corporations differently from money the business borrows from the bank

But borrowing from the trust funds is not just some paper transaction without an impact on anybody's life. Social Security and Medicare are real, tangible obligations. People get checks every month. Hospitals and doctors have to get paid. And as we've pointed out repeatedly, pretty soon the money coming into Social Security and Medicare won't be enough to cover what's promised to go out. That's when the Social Security and Medicare systems will start redeeming those Treasury bonds they've been given as IOUs. If you add the two kinds of debt together, you end up with gross debt worth 64.7 percent of GDP, which is much less reassuring.

And when this intragovernmental debt stops being a bookkeeping maneuver and becomes a real bill, who pays for it? You do. The government will have to come up with the cash to pay back the trust funds—and you're the one it gets the cash from.

CHAPTER 7

If You Think Social Security Is Bad, Wait Till You Meet Medicare

As for me, except for an occasional heart attack, I feel as young as I ever did.

—Robert Benchley, American
actor and writer (1889–1945)

When it comes to tackling the country's long-term financial problems, we often talk about having to "fix" Social Security and Medicare in one breath, as if the two programs were matching bookends—little public policy identical twins.

From the typical American's point of view, the two do have a lot in common. You have taxes taken out of your salary to pay for them. Almost everyone over a certain age benefits from them. They are very popular. They are both "entitlement programs," which means that Congress doesn't debate and vote on them every year the way they do on the defense and education budgets. When people who've paid into the systems

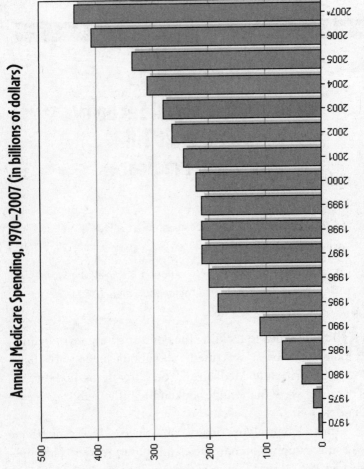

Annual Medicare Spending, 1970–2007 (in billions of dollars)

Medicare spending has risen dramatically over the last few decades because the number of older people has increased and because health care spending has gone up overall. *Source: Centers for Medicare and Medicaid Services* reach the eligible age, they are "entitled" to get the benefits. This arrangement means older Americans don't have to watch yearly congressional debates on whether to "reauthorize" the programs they depend on (complete with all the partisanship, lobbying, and other accoutrements congressional reauthorization usually entails). That's probably a merciful thing, but it also makes these programs tough to reexamine even when there are good reasons for reexamining them.

But for experts and elected officials concerned about the budget and the debt, the two programs are in very, very different situations.

MEDICARE IS ALREADY RUNNING ON EMPTY

One major difference is that Medicare is in much more trouble financially. There are good reasons to worry about the long-term financing of Social Security, but the program will actually collect enough taxes to cover what it needs to pay out for nearly a decade more. It also has money in its "trust fund" to cover its costs until 2041 (although as we've explained before and we'll explain again, that's not nearly as reassuring as it sounds, given that the country has in effect "borrowed against" the trust fund and is now nearly $9 trillion in debt). Nevertheless, the country does have a small window to adopt reforms and still give individuals some time to adapt to any adjustments we have to make.

According to the Medicare trustees' 2007 report to Congress, its "financial difficulties come sooner—and are much more severe—than those confronting Social Security."[1] Just

---

[1] Social Security and Medicare Board of Trustees, Status of the Social Security and Medicare Programs: A Summary of the 2007 Annual Reports, "A Message to the Public," available at www.ssa.gov/OACT/TRSUM/trsummary.html.

what you wanted to hear, we're sure. The part of Medicare that covers hospital costs for older Americans already pays out more than it takes in from payroll taxes.[2] Making the situation even worse, Medicare's costs are going up much faster than Social Security's. According to calculations from the Congressional Budget Office, Medicare spending will increase five times faster than Social Security's in the next couple of decades.[3]

Medicare Spending by Category, 2006

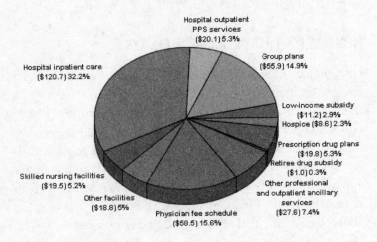

Where does the Medicare money go? Mostly for hospital care, doctors' visits, drug coverage, and other health care needs. *Source: Congressional Budget Office*

---

[2] Ibid.

[3] Testimony of Isabel Sawhill, senior fellow and director, Economic Studies, Brookings Institution, before the U.S. House Committee on the Budget, February 15, 2006, citing CBO's "The Long Term Budget Outlook," December 2005, Scenario 1.

MEDICARE'S FINANCES WILL BE HARDER TO FIX

The "fixes" for Social Security are not especially pleasant, but they probably won't put most of us in the poorhouse. No one likes the idea of raising payroll taxes, for example, and there are lots of reasons why raising Social Security taxes may or may not be a good idea (see chapters 8 and 9). But the numbers are still tethered to earth. Raising Social Security payroll taxes by about 16 percent, cutting benefits by about 13 percent, or doing some combination of the two would keep the system solvent for another seventy-five years.[4] To do the same for Medicare would require a *122 percent* increase in payroll taxes or a *51 percent* reduction in Medicare expenses, or some combination.[5]

As of right now, Social Security doesn't even collect taxes on salaries over $102,000,[6] so one widely discussed idea is to "raise the ceiling"—that is, collect taxes on the entire salaries of higher wage earners (or more of them, perhaps). In contrast, the Medicare system just has much less room to maneuver. Medicare already collects taxes on nearly everything everyone earns, so there's hardly any place to go there. Older people already pay monthly premiums to cover doctors' visits (Medicare Part B), premiums that increased annually by double digits from 2001 through 2006 (the increase for 2007 was 5.6 percent).[7] And as of 2007, most seniors were

---

[4] Social Security and Medicare Board of Trustees, "Message to the Public."

[5] Ibid.

[6] Social Security Online, Automatic Increases, Contribution and Benefit Base, available at www.ssa.gov/OACT/COLA/cbb.html#Series, accessed August 17, 2007.

[7] "Medicare Funding Change Aids Majority," editorial, *Denver Post*, September 18, 2006.

paying about $93 a month for this insurance.[8] But the system has also raised premiums substantially for higher-income recipients; these can go over $150 a month, depending on how much income the person has.[9] We're not trying to suggest that raising Medicare taxes, premiums, and fees should be off the table, and the increases on higher-income people affect only about 4 percent of recipients.[10] But politically speaking, the system has already been there and done that.

MEDICARE'S FUTURE COSTS ARE NOT AS PREDICTABLE

Figuring out how much money Social Security will need to pay future recipients is not as easy as you might think. The calculations can change, depending on how healthy the U.S. economy is (when people aren't working, they don't pay Social Security taxes), how many older people the country will have, how long they decide to work, and how long they live and collect benefits. Luckily, there are actuaries who love doing just this kind of calculating, and, in a sense, it's pretty much a closed question: a certain amount of money will come in, and the system will pay out a certain amount in benefits according to its rules and regulations.

Social Security pays people benefits that have been calculated and spelled out in advance in that nice little green-and-white newsletter you get in the mail every year (Be sure to check out your own Social Security statement when it

---

[8] Social Security and Medicare Board of Trustees, "Message to the Public."

[9] Robert Pear, "Government Sets Higher Medicare Rates and New Surcharge," *New York Times*, September 13, 2006; Tony Pugh, "Medicare Premiums to Increase," *Myrtle Beach Sun-News*, September 13, 2006.

[10] Centers for Medicare and Medicaid Services, "Medicare Premiums and Deductibles for 2007," September 12, 2006.

comes). Social Security doesn't pay you more if you hit hard times and need more money. That's your problem, and your kids' problem. Some people do try to cheat Social Security, of course, but it's not a program bedeviled by waste and fraud.

Medicare is a much more confusing problem. Since it pays for older people's hospital bills, surgeries, medical tests, medications, and other health care expenses, its costs are essentially linked to the country's overall health care system. And as we all read nearly every day (and as some of us are unfortunate enough to find out personally), health care costs are skyrocketing. That means that Medicare spending is not as foreseeable or easily controllable as Social Security spending. There are lots of different factors that can drive up costs.[11]

Like Social Security, Medicare's costs depend on how many older people the country has and how long they live. But its costs also depend on how sick people are. For example, the Alzheimer's Association recently raised its estimate of the number of Americans with the disease to more than 5 million.[12] Since caring for a patient with dementia costs Medicare about three times as much as caring for a more typical beneficiary, Medicare's costs are expected to rise as well.[13] It's a dreadful disease and a costly one as well.

There's also new research from the University of Michigan Health and Retirement Study (sponsored by the National Institute of Aging), which surveys more than twenty thousand middle-aged Americans, suggesting that the baby-boom generation is less healthy than the previous generation, possibly

---

[11] Congressional Budget Office, "The Long Term Budget Outlook," December 2005, available at www.cbo.gov/ftpdocs/69xx/doc6982/12-15-LongTermOutlook.pdf.

[12] Jane Gross, "Prevalence of Alzheimer's Rises 10% in 5 Years," *New York Times*, March 21, 2007.

[13] Ibid.

because of greater obesity and less exercise.[14] "If people are entering early old age in worse health, it doesn't bode well for society. It's quite worrying," said Richard M. Suzman of the National Institute on Aging.[15]

But Medicare's costs can go up with good news, too. If someone discovers a new treatment that will eliminate the pain and disability of arthritis, Medicare will add it to the list of covered services. It will be a wonderful thing, but it will also drive up costs. Then there's the question of whether all the services Medicare patients receive are actually medically necessary, or whether some large portion of money is being wasted. Sometimes these questions are judgment calls—should Medicare cover cataract surgery when the patient could get by with glasses? Should Medicare drug plans cover Viagra, or, since human beings actually can live without sex (yes, there is empirical evidence showing this), should Viagra be considered a personal expense? As of 2007, Medicare drug plans can cover Viagra if the patient takes it for something like "pulmonary hypertension" but not for erectile dysfunction.[16]

Sometimes the issue is outright dishonesty, and Medicare fraud horror stories are not uncommon: The government paid over $2 million for ankle braces for people who have had a foot amputated; schemers operating a sham medical supply company used the money to buy a Rolls-Royce Phantom for themselves.[17] The Government Accountability

---

[14] Rob Stein, "Baby Boomers Appear to Be Less Healthy than Parents," *Washington Post*, April 20, 2007.

[15] Quoted in ibid.

[16] Patricia Anstett, "Medicare Limits Sex Drug Coverage," *Detroit Free Press*, January 27, 2007.

[17] Carrie Johnson, "Medicare's $869 Air Mattress Bill; Government Arrests 38 as It Cracks Down on Health Care Fraud," *Washington Post*, May 10, 2007.

Office has called on Medicare administrators to do more to tackle fraud, as have some members of Congress, and the FBI has joined in on the hunt as part of a Medicare Fraud Strike Force.[18] Still, like Willie Sutton, who robbed banks because "that's where the money is," thieves are likely to see Medicare as a promising target for some time to come.

THE ANSWERS TO MEDICARE'S PROBLEMS AREN'T AS CLEAR

We might not like some of the answers about how to fix Social Security, and there is a lot of "yes, it will, no it won't" bickering over whether private accounts would make the system more or less financially stable. But compared to the debate over Medicare, the debate over Social Security is a model of mathematical clarity. We could raise taxes, cut benefits, do some of both, and/or start to plan for a private accounts option in the future. The major questions we need to talk about are value questions—what solutions we think are fairest, what kind of system we really want.

But with Medicare, we could raise taxes and fees dramatically, and if health care costs continue to rise the way they have lately, the system could still be in trouble. You might be able to cut benefits some, but when older people are fragile, sick, and dying, the country is going to want to pay for whatever will help them. The big wrinkle here is that we haven't yet come up with foolproof ways to hold down health care costs without harming patients or thwarting new medical breakthroughs. In the last couple of decades, experts have pinned their cost-cutting hopes on HMOs, preferred provider plans, more preventive care, and other ideas, but health care costs have continued to climb. The well-respected (and very quotable) budget expert Douglas

---

[18] Ibid.

Holtz-Eakin has been candid on this one: "Until we diagnose it, we can't fix it very well"[19]

THE MEDICARE DEBATE HAS BARELY STARTED

No matter what you think of President Bush's idea about starting private accounts in Social Security, he probably did the country a big favor by getting the Social Security issue out on the table. Soon there are going to be a lot more people retiring and fewer workers paying into the system. Most of us understand that we're involved in a national debate about how to pay for Social Security, and that we're probably talking about cutbacks of some sort. In fact, the system has already been slowly pushing back the age of retirement. Those of us who can are rapidly throwing money into other kinds of retirement accounts to supplement our Social Security benefits. We've basically got the picture. The system can't continue on like this, and we've got to decide what to do.

But the debate over how to handle Medicare's financial problems and keep the system solvent is barely out of the starting gate. Whether it's in Congress, on the campaign trail, in newspapers and newscasts, or at the typical family kitchen table, most of the time we talk about Medicare, we're talking about what else the program should cover— not how to contain its costs to government and taxpayers. Take, for example, the addition of prescription drug coverage to Medicare in 2003. President Clinton originally proposed the idea; Democratic candidate Al Gore talked about it, too. President Bush and the Republicans who controlled Congress in 2003 made it a reality. The law is expected to

---

[19] Transcript of Douglas Holtz-Eakin presentation at the Maxwell School/Public Agenda Policy Breakfast, April 19, 2006, available from Public Agenda.

add $518 billion (that's *billion* with a *b*) to Medicare's costs between 2007 and 2013.[20] Yet when the plan was up for discussion, the potential impact on the nation's finances barely figured in the debate. Whatever its specific merits or problems, the legislation responded to a broad public consensus that we wanted to help seniors with their drug expenses.

Still, it does seem reasonable to ask whether we would have done it in quite the same way if we had been talking about the costs up front. Columnist Robert Samuelson is no fan of the drug plan. He points out that about "three-quarters of Medicare recipients already had drug coverage. The poorest had it through Medicaid; many retirees had it from their former employers and some had it through Medicare managed-care plans or private insurance policies they purchased."[21] Yet the country passed a new and universal benefit—yes, Donald Trump will be eligible, too—without really talking much about the cost. David Walker, the U.S. comptroller general—essentially the country's accountant—called the prescription drug legislation "probably the most fiscally irresponsible piece of legislation since the 1960s."[22] According to Walker, the nation "promised way more than we can afford . . . we're not being realistic."[23]

---

[20] Congressional Budget Office, "The Budget and Economic Outlook: Fiscal Years 2008 to 2017," January 2007, available at www.cbo.gov/ftpdocs/77xx/doc7731/01-24-BudgetOutlook.pdf.

[21] Robert Samuelson, "Benefit Disaster," *Washington Post,* November 23, 2005.

[22] Interview with David Walker by Steve Croft, *60 Minutes,* CBS News, March 2, 2007.

[23] Ibid.

CHAPTER 8

Glib Answers to a Tough Problem

We won't acknowledge choices, contradictions, unpalatable facts. So, many problems persist for years. Throwing the bums out is a venerable tradition, but what if the ultimate bums are us?

—*Robert Samuelson*, Washington Post, *November 1, 2006*

The choices the country faces if we just ignore the problems facing Social Security and Medicare are so devastating that you might assume the country's leadership is talking day and night about how to avoid them. Some experts are trying to get leaders to tackle this problem more seriously (see "The Heroes of the Revolution" on page 226 to read about some who've gone to the mat trying to address the country's budget problems). But too many elected officials, not to mention talking heads and radio hosts, seem determined to stall off genuine debate with sloganeering and whining.

In writing this book, we've come to believe that there are two realities the country needs to agree on so we can get down to work. They deserve to be in big bold print, so there's no mistaking them.

One is that **the American people are not going to eliminate Social Security and Medicare,** nor are we going

Social Security and Poverty

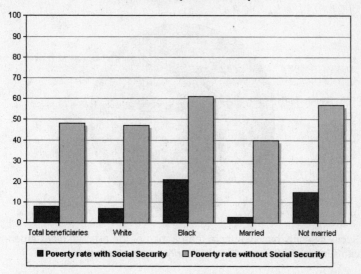

There's no question Social Security makes a difference to the people who get it. Poverty among older Americans would increase considerably without Social Security. *Source: Social Security Administration*

to transform them so they're completely unrecognizable. These programs are too popular, and too many people are counting on them.[1]

The other is that **Social Security and Medicare have to change, and they may have to change substantially.** The nation cannot afford to keep them exactly and precisely the way they are right now. Nearly every expert we can find accepts both these points, as do millions of ordinary Americans. But you would never know it from some of the

[1] See, for example, Pew Research Center, "Bush Failing in Social Security Push; AARP, Greenspan Most Trusted on Social Security," March 2, 2005; 79 percent of the public says that Social Security has had a "good" impact on the country.

Sources of Income for Older Americans

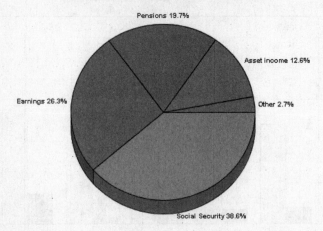

Pensions 19.7%

Asset income 12.6%

Earnings 26.3%

Other 2.7%

Social Security 38.6%

Social Security provides more than a third of the average older American's income. *Source: Social Security Administration*

screaming and daydreaming that comes in via sound bites and the blogosphere.

SIX STATEMENTS THAT ARE NOT HELPING ANYBODY

Heard anything like these lately? Everyone's entitled of course, but in our humble opinion, these six ideas aren't getting us anywhere.

No. 1: "Social Security and Medicare just aren't on the table. These benefits were promised to people." In 1940, when Social Security first started paying out benefits, life expectancy was about sixty-three years, compared to nearly seventy-eight years today.[2] Since life expectancy has increased consid-

---

[2] National Vital Statistics Reports, vol. 54, no. 14, Table 11, "Life Expectancy by Age, Race, and Sex," available at www.cdc.gov/nchs/data/nvsr/nvsr54/nvsr54_14.pdf.

erably and since the program faces financial difficulties, there's every reason to take a new look at how Social Security works. As for Medicare, health care costs are rising so rapidly and the program's financial prospects are so gruesome that there's really no choice but to change it. Surveys show that most Americans want to honor the core values of these programs—joining together to help all of us plan for retirement, making sure low-income Americans aren't destitute when they can't work anymore, ensuring that older Americans get the health care they need. Most Americans also think these programs have accomplished a lot of good, so not too many are going to go along with a slash-and-burn do-over. But that doesn't mean that Social Security and Medicare are untouchable. In fact, anyone who cares about saving them needs to open his or her mind to changes that will make them affordable again.

No. 2: "Let's just privatize Social Security right now." The idea of letting people put their Social Security payroll taxes into private accounts and invest the money as they see fit is an attractive idea to some Americans, and in the next chapter we include a proposal for gradually changing the system so people's accounts are more like 401(k)s. It's one option among many. But if you've been reading along with us, you know the downside to trying to do the private-accounts solution in a big way any time soon—and it's a whopper. We've said it before, but it's got to be said again: Social Security is a pay-as-you-go program. If people working now put all or even some of their Social Security taxes into private accounts, the country wouldn't have the money to pay for benefits for people who are old now.

And remember, Social Security is not just some generous, charitable gesture on the part of younger Americans. The older generation paid their Social Security taxes when they were working. President Bush has proposed a version of this idea in which younger workers could put a small percentage of their Social Security taxes into private accounts and invest the money. The problem is covering the hole that's left when the money moves into private accounts. The Congressional Budget Office suggests the hole would be about $270 billion just between 2011 and 2017. The Bush administration (which thinks the idea will be very popular with younger workers) projects a hole of more than $600 billion in the same time period[3]—not numbers to be taken lightly. Even some who support the private accounts concept—former Fed chairman Alan Greenspan, for example—warn against moving to such a plan too quickly. "If you're going to move to private accounts, which I approve of," Greenspan testified to Congress in 2005, "I think you have to do it in a cautious, gradual way."[4]

No 3: "Social Security? What a rip-off! I could do better keeping the money and investing it myself." This is a variation on the theme above, and whether it's actually true or not may depend on where you sit on the income scale (and how good an investor you are). If you're a lower-income worker, Social Security is not such a bad financial deal. You'll actually get a better return on your money than someone who earned higher wages and paid higher payroll taxes during his or her working years.[5] If you're higher income and a good investor,

[3] Congressional Budget Office, "An Analysis of the President's Budgetary Proposals for Fiscal Year 2008," March 2007, available at www.cbo.gov/ftpdocs/78xx/doc7878/03-21-PresidentsBudget.pdf.

[4] Edmund L. Andrews and Richard W. Stevenson, "Greenspan Backs Idea of Accounts for Retirement," *New York Times,* February 17, 2005.

[5] Congressional Budget Office, "Is Social Security Progressive?" December 15, 2006, available at www.cbo.gov/ftpdocs/77xx/doc7705/12-15-Progressivity-SS.pdf.

you might well do better with the money, but this is just one piece of the puzzle. Your taxes pay for people who are elderly now. Unless we want to see huge increases in poverty and illness among older Americans and near chaos in the health care system, we need to figure out how to make these programs financially sound for the people who need them. Moving gradually to a system in which people manage their own money in private accounts may or may not be a good idea, but if you're a responsible, compassionate person, you have to admit that this debate involves more than your own financial acumen.

No. 4: "The government can't do anything right. Look what a mess they've made of this. They ought to just get out of it." There's a segment of the population that subscribes to one single political thought—don't trust the government to do anything, period. The government certainly doesn't deserve very good marks for the way it's handled Social Security and Medicare funding so far. "D minus" might be overly generous. On the other hand, Social Security probably looks reasonably good to Enron employees who lost their retirement savings when they listened to the executive-office swindlers they worked for, or airline employees whose pensions got whacked when their companies went into bankruptcy protection, or other Americans whose private retirement plans aren't as rosy as they once thought. It's easy to knock Social Security, but not that many Americans really can do without it.[6] As for Medicare, well, good luck with that—just try to find health insurance on your own when you're old and sick.

---

[6] In 2001, the Social Security Administration calculated that 48 percent of recipients would fall below the poverty line without their monthly check. See www.publicagenda.org/issues/factfiles_detail.cfm?issue_type=ss&list=9.

No. 5: "Not to worry. The Social Security trust fund doesn't run out of money until 2041." Well, not exactly. We talked a little about this in the last chapter, but it's worth going over again, because it *is* confusing, and to those of us who aren't Beltway insiders, it's a little mind-boggling as well. The baby-boom generation is huge compared to the generations before it and after it, so while the boomers were (and are) working, Social Security brings in more money than it needs to pay out benefits, and that extra money is supposed to be in a "trust fund." Now, most of us hearing the words *trust fund* might think that the extra money is being invested or socked away to earn interest, that it's sitting somewhere lovingly protected by folks wearing green eyeshades working in offices with a lot of mahogany furniture. But that's just not the case. The government continually and repeatedly borrows from the trust fund to cover other government expenses. The trust fund does hold a lot of "special-issue Treasury securities,"[7] and it is entitled to redeem them. We all know about the "full faith and credit" of the U.S. government, and that's reassuring. But to pay back the money, the government will have to raise taxes, cut other government spending, or borrow even more money (and over time this is going to be a huge, huge amount of money). So don't relax—definitely not. By the way, there is a Medicare trust fund as well, but since it's going into the red as we speak, no one is even suggesting relaxing on that one.

No. 6: "The government spends too much on the elderly and not enough on kids." It's because old people vote, and children can't. This one is a party favorite in some D.C. public policy circles. If

---

[7] National Academy of Social Insurance Sourcebook, "Where Do Social Security Taxes Go?" available at www.nasi.org/publications3901/publications_show.htm?slide.

you hear someone calling for more "intergenerational equity," this is basically what's bothering them. Essentially, it's a debate over who gets more, and who needs more, support from the government—the elderly or kids? According to a study from the Urban Institute, about 46 percent of domestic federal spending goes for programs that mainly serve the elderly—Social Security, Medicare, and Medicaid, which pays for nursing home care as well as care for the poor—while just 15 percent is spent on children.[8] The bulk of state and local taxes are spent on children (for public schools, mainly), but even then, the kids still don't catch up. Of course, most kids live at home with their parents, so they don't have to pay for food and housing the way most elderly people do, and children don't generally have serious health care problems the way most elderly people do. On the other side, a higher percentage of children live in poverty than elderly people. You can see how the back-and-forth on this could go on and keep on going. What bothers us here is the groupthink and the idea of setting this up as a fight between generations. People who like kids over here. People who like older folk over there. Choose your side. To our way of thinking, this just doesn't seem like a very productive path. In fact, one of the major reasons to act now to get Social Security and Medicare on a sound footing is to make sure that we have money left to do what we need to do for our kids, especially the poorest of them.

DON'T REACH FOR THE EASY BUTTON

Our point here is that as a country, we need to start talking seriously and realistically about what to do and cut it with

---

[8] Adam Carasso, C. Eugene Steuerle, and Gillian Reynolds, "Kids' Share 2007," Urban Institute, March 15, 2007, available at www.urban .org/publications/411432.html.

the manipulating and hyperventilating. These are complicated issues, and we need to listen to each other and try to get the best thinking on what will work and what won't. We're not going to get there by falling for "this is all simple if you'll just do it my way" bloviating.

In the next chapter, we'll run down some of the real choices you need to start thinking about. There are ideas for modifying the way Social Security and Medicare operate and ideas for changing the way we pay for them. They probably aren't as fun to read about as what passes for policy discussion on some cable news shows and blogs, and chances are you're not really going to be crazy about any of them. Unfortunately, there just doesn't seem to be an "easy button" for this one. But after you think about it, you can probably live with enough of them to make a real difference, enough to start getting the country back on the right financial track.

CHAPTER 9

Do We Have to Throw Granny out on the Street?

The longer we wait, the more severe, the more draconian, the more difficult the adjustment is going to be. I think the right time to start is about 10 years ago.

—*Ben S. Bernanke, Federal Reserve chairman, January 2007*

The choices for fixing Social Security and Medicare are not much fun, which is why politicians don't talk about them much. Fixing the problem means cutting back benefits and/or raising taxes in some form or another, so this is not a happy topic of conversation for people whose jobs depend on being popular. Surveys routinely show that most Americans want to save Social Security and Medicare, but reject most workable solutions.[1]

Luckily, your authors are *not* running for office, so we're presenting for your consideration (in our best Rod Serling

---

[1] See Public Agenda Online—Issue Guides: Social Security, at www .publicagenda.org.

Social Security is projected to begin paying more in benefits than it collects in taxes in 2017.
Credit: Franklin D. Roosevelt Library

voice) fourteen different ways to approach the problem. It's not a complete list, but it will get you started. And be forewarned, reading this chapter will probably take some effort and concentration. These proposals are the wheat germ, brussels sprouts, and plain yogurt of budget policy (and we know that you'd rather have the nachos).

What's more, you can't just choose one idea out of our fourteen and leave it at that. The country will have to make a number of changes on different fronts to really make a dent in the problem. And, obviously, we'll need to consider reforms to tackle both Social Security and Medicare. And while the choices in this chapter focus directly on entitlements, the country will have to make other decisions that will affect what kind of debt and economy we leave to the next generation. You can check out chapter 14 for some choices on taxes that we'll also need to be considering.

As of now, the experts haven't done all the number crunching, computer modeling, and task force hobnobbing they need to do for us to tell you authoritatively exactly how all of these ideas would work and exactly how much each one would save in comparison to the others. And when the debate really gets going in Washington, there will be lots of variants and adaptations of what we describe here. That's just how life works in the policy-making world. So you might want to think of these as "starter ideas," not full-fledged policy proposals. Even so, approaches like these are being talked about in government offices, think tanks, editorial board meetings, and policy conferences every day. There's no reason you shouldn't be in on this.

If you start getting a little depressed as you glance down the list, remember the dreadful alternatives facing us if the country doesn't act on the Social Security/Medicare issue (they're back in chapter 6, if you need a shot of courage). The worst choice of all is to do nothing. A list of publications and Web sites with more on these ideas and some we couldn't cover here is in the appendix. Here goes.

1. Move the retirement age up to seventy. Polls show that this is one of the least popular ways to shore up Social Security, but it would save money.[2] Americans are living longer, and people in their seventies are barely considered old now (at least not by people in their forties and fifties), so there's a commonsense argument

---

[2] Public Agenda Online, Social Security: Bills and Proposals, "Majorities Oppose Proposals to Raise the Retirement Age, Reduce Benefits, or Increase Taxes, but Support Limiting Benefits and Increasing Taxes for the Wealthy to Secure the Future of Social Security," available at www.publicagenda.org/issues/major_proposals_detail.cfm?issue_type =ss&list=5.

for moving the retirement age back. In fact, Social Security has already started edging in this direction. Despite everything you hear about sixty-five being "the age of retirement," people born between 1943 and 1954 have to wait until they're sixty-six to collect full Social Security benefits. People born in 1960 or later have to wait until they're sixty-seven.[3] But most of us don't even wait that long. Sixty-five, sixty-six, and sixty-seven may be the official ages of eligibility for full retirement benefits, but Social Security lets recipients collect benefits at reduced rates at sixty-two, so a lot of people choose that option.[4] According to the Employee Benefit Research Institute, the *average* age of retirement is sixty-two.[5] Only about three in ten older Americans retire at age sixty-five or older.[6] Most of us probably could work a few years longer without much trouble, although this might be tougher for people doing manual labor. Some jobs just get harder as you get older. To allow people to plan and adjust, Congress would probably phase in the older age over a number of years. Even done gradually, however, the Congressional Budget Office predicts that doing something like this could "reduce Social Security outlays by 14 percent"—a significant amount.[7]

2. Make people pay Social Security taxes on more of what they earn. As of 2008, Social Security taxes for you (and your employer) stop

---

[3] Social Security Online—Find Your Retirement Age, at www.ssa.gov/retirechartred.htm.

[4] Ibid.

[5] Employee Benefit Research Institute, Fast Facts, "When Do Workers Plan to Retire vs. Actually Doing So," available at www.ebri.org/pdf/publications/facts/fastfacts/fastfact050107.pdf.

[6] Ibid.

[7] Congressional Budget Office, "Budget Options," February 2007, available at www.cbo.gov/ftpdcs/78xx/doc7821/02-23-BudgetOptions.pdf.

when your salary reaches $102,000.[8] If you're lucky enough to earn a salary in this range, you have probably enjoyed seeing your paycheck get quite a bit fatter the last few months or weeks of the year because Social Security taxes aren't being deducted anymore. Taxing more of people's salaries would generate more revenue for the system, and people who like this option point out that it would hit higher-income people hardest. The CBO looked at one proposal that would tax salaries up to $250,000 and calculated that it would produce nearly $285 billion between 2008 and 2012.

Currently, employee taxes are matched by employers, so if the new law continued this arrangement, this would also be a big tax increase on business. It's always nice to make "business" pay for things, but as we've mentioned before, "business" isn't the only one who ends up paying. When businesses are hit with higher taxes, they also tend to hold down salaries (we all like raises), lay people off (which hurts both individuals and communities), reduce investment in new equipment and facilities (which is bad for the economy) and/or pass the expense on to consumers (that's you and me, in case you had forgotten).

3. Change the way Social Security benefits are calculated. If you visit the Web site for the Congressional Budget Office, you can read about a proposal to hold down Social Security costs by "constrain[ing] the increase in initial benefits."[9] It's all about changing the formula for how Social Security calculates what your benefits will be when you retire. Don't worry if you find the description a little rough going. Even Social Security

---

[8] Social Security Online, Frequently Asked Questions, Benefits, "What Are the Tax, Benefit and Earning (COLA) Amounts for 2007?" (www.ssa.gov), accessed August 11, 2007.

[9] Congressional Budget Office, "Budget Options."

admits that "the benefit computation is complex."[10] And the key question is not how well you grasp the formula, but how you feel about the pros and cons of this option compared to others. As Social Security describes it, your "benefits are based on earnings averaged over most of a worker's lifetime" and "your actual earnings are first adjusted or 'indexed' to account for changes in average wages since the year the earnings were received."[11] The proposal would change the formula to base it on how much *prices* have increased, not how much *wages* have increased.[12] (Yes, there are people in Washington who have to think about these details.) The bottom line is that people retiring after the change would get somewhat lower benefits than under current rules.

For example, according to the CBO, someone retiring in 2030 would get 24 percent less under the new formula than under the current one. Someone retiring in 2050 would get 40 percent less.[13] That's a big cut, but the changes would be gradual—giving everyone time to plan—and benefits would keep up with inflation because they would be tied to price increases. But in the end, people retiring in the future wouldn't get as much as people who have retired up to now.[14] Still, according to the accounting wizards at the CBO, this one idea by itself would reduce Social Security's long-term financial problems by more than 30 percent by 2050.[15]

---

[10] Social Security Online Frequently Asked Questions, Computation of Benefits. "How Are My Retirement Benefits Calculated?" (www .ssa.gov), accessed August 18, 2007.

[11] If you like math, it's at www.cbo.gov/publications/collections/social security.cfm.

[12] Congressional Budget Office, "Budget Options."

[13] Ibid.

[14] Ibid.

[15] Ibid.

The downside? Reducing benefits, even gradually like this, is not popular. And it may be even less popular if it seems to be the result of some complicated mathematical calculations that nearly all of us have trouble understanding. The plan could be adjusted to provide more protection for low-income people, but middle-class people would probably have to save considerably more on their own to be as well off as retirees today.

4. Reduce Social Security and Medicare benefits for wealthier Americans.

Social Security and Medicare are available to every American who pays into the programs, no matter how high his income or how much money he has in the bank. The systems were organized that way so they wouldn't be seen as "welfare programs" and would have broad support. There's also the argument that if you pay money into these systems while you are working, you should get something back even if you are fairly well-off. Given the big crunch to come, however, some say it's time to rethink this, and polls suggest that the general idea of cutting back benefits for wealthier Americans has more potential support than other approaches.[16] This could play out in a lot of ways. There could be formulas allowing lower- and middle-income people to get more from Social Security and Medicare, while higher-income folk get less. You could redesign both Social Security and Medicare so that they are mainly intended for low- and middle-income people—the programs just wouldn't cover you if your retirement income from other sources was more than, say, $100,000 a year. And there, as Shakespeare put it, is the rub. How do you define "wealthy" and how do you define

---

[16] Gallup Organization for CNN/USA Today, February 2005, available at www.publicagenda.org. See Issues/Social Security/Bills and Proposals.

"middle class"? Expect to hear a lot of talk (and argument) about this approach in years to come.

5. Change the way Social Security cost-of-living increases are calculated.

Whether you're working now or you're retired and collecting a pension or Social Security, you can imagine what would happen over time if your income never went up. You might be fine at first, but after a number of years, your standard of living would really slide. That's why Social Security (and many other pensions and employee contracts) have cost-of-living increases. Social Security uses a version of the Consumer Price Index (CPI-W) to calculate COLAs (cost of living allowances) for recipients, so people on Social Security received a 4.1 percent increase in 2006 and a 3.3 percent increase in 2007.[17] Some experts say the current formula exaggerates the actual increase in people's living costs,[18] and they want Social Security to use a less-generous one. Much of the argument here is over whether the formula used now is overly generous.[19] Those who defend the current system point out that older people spend more of their income on health care, and that health care costs are rising faster than most other costs. They also argue that older people can't take on extra hours or change jobs to earn more the way a younger person could. People who reach their full retirement age (sixty-five, sixty-six, or sixty-seven, depending on when you were born) can collect

---

[17] Social Security Online, Frequently Asked Questions, Benefits, "Will My Benefit Amount Be the Same for the Rest of My Life?" (www.ssa .gov), accessed August 11, 2007.

[18] Congressional Budget Office, "Budget Options."

[19] The Bureau of Labor Statistics at the U.S. Department of Labor has a good description of how the CPI is calculated and what it includes at www.bls.gov/cpi/cpifaq.htm#Question_1.

Social Security plus whatever they can earn on top of it,[20] but this just may not be an option for people over a certain age. Like so much else in this debate, there are two sides to the story—if not more.

6. Privatize Social Security, but very, very slowly. It's controversial, but the idea of letting workers put their Social Security tax payments into private accounts like 401(k) plans does have some appealing aspects. One is that people would actually own and control their own accounts, so they wouldn't have to rely on the government to have the money available when they retire. Another plus is that if people had money left when they died, they could leave it to their spouse or children or pets or some other good cause. The snag, and it's a big one, is that the government needs the money workers pay into Social Security *now* to pay benefits to people who are retired *now*. If workers put that money into a 401(k)-type plan instead, there would be a big hole. To fill it, the country would need to raise taxes, cut other government programs, borrow more money, or reduce benefits for people who are retired—none of which are very attractive prospects. Still, some privatization advocates think the idea is worth the costs if we make the changeover very, very gradually. In seventy-five years or so, these experts argue, we could have a system based mainly on private accounts that would be better than what we have now. Of course, others say the idea of taking money out of Social Security just when the boomers are beginning to retire is irresponsible, to say the least. They also worry whether people will make good investment decisions with their private accounts and what will happen to people who retire when the stock market is down. Many just

---

[20] Social Security Online—Retirement Planner, available at www .ssa.gov/retire2/whileworking.htm.

don't see why we would move away from one of the most popular and effective government programs ever created for something that may or may not work as planned.

Medicare was one of the centerpieces of President Lyndon B. Johnson's "Great Society" when it was signed into law on July 30, 1965. *Credit: LBJ Library*

7. Up the eligibility age for Medicare. As we explained up top, Social Security has already raised its eligibility age to sixty-seven for those born in 1960 or later, but people can still go on Medicare at sixty-five. Having the same rules for Medicare and Social Security seems like a good way to neaten up the system, and since Americans are living so much longer, it seems logical for both of these programs to kick in a bit later, especially if we decide to move the age of retirement back further. Obviously, if people start collecting Medicare benefits later, the program will cost less. One option explored by the CBO would cut Medicare costs by about 10 percent by 2050.

The downside? This is definitely not a winning campaign slogan.

Since people in this age range would still need insurance, employers might be hit with higher insurance premiums for older employees. Less-scrupulous bosses might try to ease older workers out of their jobs to hold down their own insurance bills.[21] For people who don't get insurance at work, finding it on their own once they're past fifty-five or so (and more likely to have high blood pressure, brittle bones, and other common ailments of aging) could be expensive and difficult.

8. Increase the fees and deductibles older Americans pay for Medicare. Anyone on Medicare (or who has a parent or grandparent on Medicare) knows that it's hardly a free program. There are premiums and fees and deductibles galore, and they are confusing, to say the least. There are deductibles for each "spell" of illness; copayments for extended hospital care; higher deductibles for outpatient care; no deductibles at all for some things.[22] Many experts say the country needs to revamp this whole aspect of Medicare, both to make it less confusing and to save money. In its winning way with words, the CBO refers to this idea as a proposal to "modify Medicare's cost-sharing requirements." It might go something like this: People on Medicare would pay the first $500 of their health care costs; then they would pay 20 percent of their expenses up to a cap of $5,000 per year.[23] According to the CBO, a plan like this would save over $11 billion over five years, and would actually be better for people who are extremely ill with very high expenses. Unfortunately, $11 billion is a pretty small amount (amazing, isn't it?) given

---

[21] Congressional Budget Office, "Budget Options."

[22] Ibid.

[23] Ibid.

Medicare's overall costs, about $330 billion in 2006 and projected to be more than $700 billion by 2017.[24] And CBO is subdued about the degree to which higher premiums and co-pays can reshape Medicare's long-term outlook. Still, a lot of experts say Medicare needs to have some fees and co-pays to discourage patients from using services that aren't really necessary (or their doctors from ordering them).[25] All in all, it's likely that increased fees of some sort will be part of Medicare reform as we go forward. In fact, Medicare has already moved in this direction, increasing some premiums for higher-income beneficiaries.

9. Rethink Medicare drug coverage. When it became the law of the land, most of the headlines about the 2003 Medicare drug plan talked about how confusing it was. But the plan had another serious drawback—it was an expensive addition to the federal budget. The latest estimates predict it will cost $518 billion between 2007 and 2013,[26] and some budget experts say we really need to look at how to reduce the program's overall costs to taxpayers. One idea is limiting the plan to low-income seniors or perhaps low- and middle-income seniors. Right now, Oprah will be eligible when she retires (provided she's been paying Social Security and Medicare taxes). Others want Medicare to bargain with drug companies for lower prices the way the VA does. The cur-

---

[24] Congressional Budget Office, "Fact Sheet for CBO's March 2007 Baseline: Medicare," available at www.cbo.gov/budget/factsheets/2007b/medicare.pdf.

[25] Congressional Budget Office, "The Long-Term Outlook for Medicare and Medicaid," *The Long Term Budget Outlook*, December 2005 (www.cbo.gov/ftpdocs/69xx/doc6982/12-15-LongTermOutlook.pdf).

[26] "Congressional Budget Office, "The Budget and Economic Outlook: Fiscal Years 2008 to 2017," January 2007, available at www.cbo.gov/ftpdocs/77xx/doc7731/01-24-BudgetOutlook.pdf.

rent law relies on competition among different insurance companies to keep prices down). Others want to repeal the law's tax incentives (or giveaways, depending on your point of view) for health care companies (to participate) and for employers (to keep drug coverage for their retirees).

10. Run Medicare more like the VA. According to some experts, the Veterans Health Administration has already demonstrated some good ways to improve patient care *and* hold down costs. Columnist Paul Krugman described the VA as a "health-care system that has been highly successful in containing costs, yet provides excellent care."[27] A trio of physicians studied the VA and concluded that in the last decade or so, it had left behind its "tarnished reputation of bureaucracy, inefficiency, and mediocre care" to become a "model system characterized by patient-centered, high-quality, high-value health care."[28]

What did the VA do? It bargained with drug companies to get better prices than most private insurers. It set up computer systems so doctors and managers could pull up patient records quickly, coordinate care, and avoid prescription errors. It emphasized geriatric and outpatient care for older vets to keep them as healthy and independent as possible for as long as possible. In recent years, the challenges faced by the VA have changed. Previously, the veterans hospitals primarily served aging vets from World War II, Korea, and Vietnam. But the wars in Iraq and Afghanistan have

---

[27] Paul Krugman, "Health Care Confidential —Veterans' Hospitals Are Unsung in Delivering Excellence," *New York Times*, January 27, 2006.

[28] Jonathan B. Perlin, MD, PhD, Robert M. Kolodner, MD, and Robert H. Roswell, MD, "The Veterans Health Administration: Quality, Value, Accountability, and Information as Transforming Strategies for Patient Care," *American Journal of Managed Care*, November 2004.

created a new wave of seriously disabled younger vets, and the VA has been criticized for not offering the quality of care these people deserve.[29]

Despite its current problems, some say that the VA was onto something. The agency used the power of a large, centrally managed system to change the way care is delivered much the way some European-style health care systems do. In fact, the VA is essentially a government-run health care system. So, is that a good thing or a bad thing? Well, it's something you'll need to consider for yourself as the debate on how to reform Medicare heats up.

11. Redesign Medicare so older people choose among different insurance plans.

Most of us take whatever health insurance our employer decides to give us, but current and retired federal workers can choose from over two hundred different health insurance plans from a range of providers.[30] The Federal Employees Health Benefits Program has a fearsome acronym—FEHBP— but some say a plan like this might work well for Medicare. In FEHBP, the employer (the government) provides a set amount for health insurance, but how much employees have to contribute depends on which plan they choose. If they choose a more-expensive plan, they have to pay more themselves. FEHBP's costs have risen some lately, according to the GAO,

---

[29] See for example, Charles M. Sennott, "For Veterans in Rural Areas, Care Hard to Reach," *Boston Globe*, April 29, 2007, and Erika Bolstad, "Improving Rate of Vets' Survival Stresses System; Head Injuries and Post-Traumatic Stress from Combat Missions in Iraq and Afghanistan Are Presenting a New Set of Challenges Back Home," *Miami Herald*, April 6, 2007.

[30] Testimony of John E. Dicken, "Federal Employees Health Benefits Program: Premiums Continue to Rise, but Rate of Growth Has Slowed Recently," United States Government Accountability Office, May 18, 2007 (www.gao.gov/new.items/d07873t.pdf).

but so have insurance costs overall, and FEHBP's costs have not risen as quickly as those of some similar plans. Those who want Medicare reorganized this way say insurance companies would hold their costs down to attract seniors to their plan and that there would be less bureaucracy and waste. Some cite what's been happening with Medicare drug coverage—in which seniors can choose among competing plans and in which costs have been lower than expected .[31]

But opponents aren't as optimistic. For one thing, elderly Medicare recipients would have to choose among competing health insurance plans (just picture it—the let's-choose-from-a-zillion Medicare drug plans scenario reloaded). Opponents also worry this won't really save money because private insurance companies—unlike Medicare—need to make profits and pay for advertising to compete for customers. Unfortunately, there's not really much research on how something like this would work with an older, sicker population. As the CBO dryly puts it, there's not much "experience on which to base long-range estimates of the effects this approach would have on total costs or to assess its impact on beneficiaries."[32]

12. Redesign Medicare so people shop around for cost-effective health care.

Some experts want to harness the power of the consumer, not to shop for insurance policies, but to shop directly for doctors, hospitals, tests, and so on. In fact, many analysts say one of the reasons health care is so expensive is that people don't pay much of the cost themselves. Consequently, people rarely even

---

[31] "Medicare Drug Costs Are Down, *Richmond Times Dispatch*, November 29, 2006; "Prices Keep Medicare Costs Down," *Grand Rapids Press*, December 3, 2006; Kevin Freking, Associated Press, "Medicare Drug Plan Below Budget, *Star-Ledger* (Newark, N.J.), November 29, 2006.

[32] Congressional Budget Office, "The Long-Term Budget Outlook."

ask about prices, much less shop around for a less-expensive doctor or hospital. There are different ways this idea could work, but one of the most talked about is medical savings accounts. People would put money into the accounts while they were working (and get a tax deduction for it). When they retire, they would use the money to pay for routine medical expenses. Serious illness would still be covered by Medicare (or private insurance, as in choice no. 11). Advocates say this would discourage people from going to the doctor or having tests and procedures that aren't really necessary because they would be using their own savings. They also believe it would encourage the health care industry to hold down costs because patients would be asking questions and shopping around.

Critics of the "shop-around" strategy worry that seniors will avoid routine care in order to save money. It's also worth asking whether Americans are psychologically ready to go with the *cheapest* doctor, even one who's not as shameless as Dr. Nick on *The Simpsons*. The good Dr. Nick does infomercials and promises any operation for $129.95—"Hi, everybody! Call me at 1-800-DOCTORB. The 'B' is for bargain!" But then again, maybe seniors will be savvier medical consumers than anyone thinks.

13. Spend less on Medicare patients who are terminally ill and use hospice care more. It's not easy to know exactly how much Medicare spends on "heroic measures" for terminally ill patients[33]— especially since people have different definitions of what counts as "heroic" care. Still, there is very little doubt that it is billions of dollars every year. Consequently, some policy

---

[33] Daniel Altman, "How to Save Medicare? Die Sooner," *New York Times*, February 27, 2005; Mary Ann Roser, "Should Cost Be a Factor in a Case Like Emilio's, Which Exceeded $1.6 Million?" *Austin-American Statesman*, May 28, 2007.

experts are suggesting, very gingerly, that Medicare needs to cut back spending on medical care that doesn't really extend or improve patients' lives.

Right now, there are surprisingly wide variations in the care dying people receive. When doctors at Dartmouth Medical School looked at this question, they found that patients in New York spend about thirty-five days in the hospital and see thirty-five doctors in their last six months of life. In Oregon, the average is eight days and fourteen doctors.[34] Americans don't necessarily agree on which approach is better. Some worry about getting *too much* medical care near the end of life; some fear getting *too little*. In many respects, this is part of our society's larger ethical struggle about what constitutes "a good death."

No one is suggesting that Medicare would stop covering care to ease pain or make patients more comfortable, but some say the program shouldn't pay for procedures and treatments that have almost no chance of success. In many cases, they say, patients and families don't even want all this health care near the end of life—they just don't know about options like hospice care or can't stop hospitals and doctors once they switch into gear.[35]

Taking this approach would probably involve some form of "rationing" (or limiting care) in some circumstances. It might involve protecting hospitals and doctors from some lawsuits. It would probably lead to wider use of hospices. While most health care experts have long thought that hospice care is cheaper than hospital care, this isn't actually a clear-cut case. The National Hospice and Palliative Care

---

[34] Stephanie Saul, "Treatments: Need a Knee Replaced? Check Your ZIP Code," *New York Times*, June 11, 2007.

[35] See, for example, Anne Applebaum, "How We Die: Choice and Chance," *Washington Post*, March 30, 2005.

Organization (which advocates hospice care) cites research saying that hospice care saves up to $8,800 per patient, depending on the illness.[36] But a RAND Corporation study reached different conclusions, saying hospice patients cost Medicare slightly more than traditional care, although for certain types of cancer there were significant savings.[37]

But for many Americans, the question is not about money. For them, the thought of not doing everything humanly possible to extend life is unthinkable—even for people nearing their end. Nearly every one of us knows an older person who seemed like they were about to die and yet rallied back to life. Sometimes this person only lived a few more months, perhaps a year. Yet the extra time can be profoundly precious to those who love that person. Consequently, this approach is unsettling to think about, ethically and emotionally.

Still, 42 percent of Americans say they would want to stop treatment if they had a terminal illness with no hope of improvement, one that made it difficult to function in their daily lives, while 43 percent would want doctors to do every-

---

[36] "Hospice Costs Medicare Less and Patients Often Live Longer New Research Shows," press release, National Hospice and Palliative Care Organization, September 21, 2004 (www.nhpco.org/i4a/pages/Index.cfm?pageid=4343); B. Pyenson, B. S. Connor, K. Fitch, and B. Kinzbrunner, "Medicare Cost in Matched Hospice and Non-Hospice Cohorts," *Journal of Pain and Symptom Management*, September 2004.

[37] "RAND Study Finds Choosing Hospice Care Raises Medicare Costs for the Last Year of Life," press release, RAND Corporation, February 16, 2004 (http://www.rand.org/news/press.04/02.16.html); Diane E. Campbell, PhD, Joanne Lynn, MD, Tom A. Louis, PhD, and Lisa R. Shugarman, PhD, "Medicare Program Expenditures Associated with Hospice Use," *Annals of Internal Medicine*, February 17, 2004.

thing possible to save their life. [38] As a public, we seem to be evenly divided on what kind of care terminally ill patients really want and need—and whether the government should be involved in this decision at all—so maybe it's time to talk about this more. [39]

14. Pass a national VAT to help pay for Social Security and Medicare. Even if we give Social Security and Medicare quite a few nips and tucks, and even if we go back to the income tax rates we had under President Clinton, the country may still need more money coming in to make these programs affordable. That's how huge this Social Security and Medicare challenge is. That's why some experts are saying that it's time to start a kind of national sales tax called a VAT to help get our financial house in order. [40] VAT stands for **v**alue **a**dded **t**ax (If you've been shopping in Paris or Madrid, this is similar to those value-added taxes you would have been entitled to get back. If you had actually filled out the forms. And mailed them to Paris or Madrid.) We've listed some places where

---

[38]   Princeton Survey Research Associates for Pew Research Center. November 9–27, 2005: "Question: Now I'm going to describe a few medical situations that sometimes happen, and for each one, please tell me what you would want YOUR OWN DOCTOR to do, if you could make the choice. Would you tell your doctor to do EVERYTHING POSSIBLE to save your life (43%), or would you tell your doctor to STOP TREATMENT so you could die (42%)? It depends, 5%; Don't know, 10%."

[39]   For a nonpartisan look at both the ethical and financial issues around end-of-life care, visit Issues/Right to Die at www.publicagenda .org/issues.

[40]   See for example, Henry J. Aaron, William G. Gale, and Peter R. Orszag, "Meeting the Revenue Challenge," *Restoring Fiscal Sanity: How to Balance the Budget*, ed. Alice M. Rivlin and Isabel Sawhill (Brookings Institution Press, 2004).

you can find some more extensive definitions of the VAT tax,[41] but basically it's applied to products at various stages of their coming to life—when the raw materials are bought, when it's manufactured, etc.—and the cost is passed on to— you guessed it—the consumer. Some economists believe that VATs are less likely to discourage investment and entre- preneurship, although others worry that these kinds of sales taxes are tougher on low-income and working people. Oth- ers say we shouldn't be raising taxes at all.

If a VAT proposal (or even a more familiar sales tax idea) moves to the head of the class in terms of the national debate, you'll have plenty of chances to think about the pros and cons, but do keep a close eye on the details. Sometimes VATs are discussed mainly as ways to simplify the current tax system—not to bring in more money.[42] And sometimes they are proposed as a way to offer a new benefit such as universal health care coverage.[43] These aren't necessarily bad ideas, but if you care about the problem we are discussing here—the habitual red ink and the coming money crunch on Social Security and Medicare—you need to keep your eye on the bottom line. If the sales or value-added tax is intended to replace the income tax or buy something new, it won't help us get out of the financial hole we are so busily digging.

---

[41] Answers.com has definitions of the VAT tax from the *Small Business Encyclopedia, Columbia Encyclopedia,* Wickipedia, and other sources; see www.answers.com/topic/value-added-tax.

[42] See, for example, Michael J. Graetz, "100 Million Unnecessary Returns: A Fresh Start for the U.S. Tax System," *Yale Law Journal,* September 2004 (www.law.yale.edu/graetzhome/extra/YLJ_Graetz%20essay.pdf).

[43] See, for example, Ezekiel J. Emanuel & Victor R. Fuchs, "Solved! It Covers Everyone. It Cuts Costs. It Can Get through Congress. Why Universal Health Care Vouchers Is the Next Big Idea," *Washington Monthly,* June 2005.

CHAPTER 10

Waste Not, Want Not

It is a popular delusion that the government wastes vast amounts of money through inefficiency and sloth. Enormous effort and elaborate planning are required to waste this much money.

—*P. J. O'Rourke, "The Winners Go to Washington,"*
Parliament of Whores, *1991*

Donald R. Matthews is a fortunate man. He lives in a nice house in a suburban subdivision, where the federal government subsidizes his backyard.

Every year, this retired asphalt contractor from El Campo, Texas, gets $1,300 in agricultural subsidies from the federal government. Never mind that he isn't a farmer. Never mind that he never even said he wanted to be a farmer. The fact is, his house *used* to be part of a rice farm, and that's good enough for the government.

Matthews feels guilty about this, according to the *Washington Post*. He even tried to give the money back to the government, but was told the government would just give it to somebody else who lives on a former rice farm. So he gives the money to local scholarship funds. "Still, I get money I don't think I'm entitled to," he said.

The federal government has sent $1.3 billion in crop subsidies since 2000 to people who don't farm anything, the *Post* reports.[1] And if that's not mystifying enough for you, this came about because of changes to the law that were supposed to *reform* the crop subsidy program—even phase it out over time.

You can go post that angry blog entry or that furious e-mail to your member of Congress now. We'll wait.

Stories like that leave people with the impression that the United States government, that marvel of checks and balances, the inspiration for freedom fighters for generations, is nothing but an enormous pork-delivery system. And there's plenty of evidence for that view if you decide to go looking for it.

★ The Federal Emergency Management Agency (FEMA), which did such a "heck of a job" responding to Hurricane Katrina, became infamous for buying trailers that storm victims never got to use, but that wasn't all. Audits showed that the agency racked up nearly $2 billion in waste and fraud in funds allocated for Katrina victims.

★ The "Bridge to Nowhere" has a romantic ring to it, but this $223 million bridge linking a small island in Alaska to the mainland became a symbol for wasteful and surreptitiously "earmarked" federal spending. Unfortunately, there seem to be other communities in line for bridges that hardly any of us will ever use. There's a $27 million appropriation for a bridge linking Lone Star with Rimini, two towns in South Carolina that have barely two thousand people between them. Nowhere must be a popular destination these days—lots of people want to get there fast.

---

[1] "Farm Program Pays $1.3 Billion to People Who Don't Farm," *Washington Post*, July 2, 2006.

★ Between 1997 and 2003, the Pentagon bought $100 million in refundable airline tickets it didn't use, and somehow they never got around to asking for refunds. On top of it, they spent another $8 million mistakenly reimbursing employees for tickets the Pentagon had purchased itself.[2]

★ In the business world, corporate credit cards are notorious temptations for big spenders ("Lemme just put this on my expense account, baby"). So why should the government be any different? One audit found 15 percent of all the Agriculture Department credit card accounts examined were being abused, with $5.8 million spent on purchases such as Ozzy Osbourne tickets, tattoos, lingerie, and tuition at bartender school.[3]

★ If you'd like to get your blood pressure up sometime, we recommend the *Congressional Pig Book Summary* from Citizens Against Government Waste, available on Web site, www.cagw.org. In 2006, there were 9,963 projects tucked into appropriations bills that met their definition of "pork," costing $29 billion.

Infuriating, isn't it? If we could just get rid of all that pork, if we just had an honest, efficient government, then we wouldn't have budget problems. Right?

Unfortunately, no. Waste is a problem, but it's not *the* problem. We're not going to call chasing waste and fraud an "easy answer"—that does an injustice to anyone who has devoted their life to fighting city hall. It's rough work, and not for the squeamish. In fact, we think anyone who has successfully battled government waste and fraud deserves the equivalent of a national standing ovation.

---

[2] "The Top 10 Examples of Government Waste," Heritage Foundation Backgrounder, April 4, 2005.

[3] Ibid.

But the harsh fact is that if we somehow found enough people who were incorruptible, smart, and really, really cheap with other people's money to staff the entire government, we'd still have an enormous budget problem to solve. Simple honesty isn't going to be enough.

Waste *does* matter. But maybe not the way you might think.

THE TRIFECTA OF THROWING AWAY MONEY

As the examples above prove, there are lots of ways to waste government money. People are creative that way. But it's helpful to define some terms. If you're going to leave this chapter angry, you ought to know exactly what you're mad at.

There are basically three categories of government money gone wrong.

★ *Fraud and abuse,* which pretty much means stealing
★ *Waste,* which can best be summed up as bungling
★ *Pork,* which is politically motivated spending designed to keep an officeholder's supporters and constituents happy

Each happens for a different reason and demands a different solution.

Category No. 1: Fraud and Abuse—or, At Least We're Doing Better than Ghana

Outright corruption pretty much speaks for itself. But another way of characterizing corruption is when people in government do things they have absolutely no right to do. Stealing government money to buy things for yourself is a pretty clear example. Those Agriculture Department workers charging tattoos and lingerie to the government knew what they were doing was wrong. Bribes, kickbacks, bid rigging, influence peddling, and all the rest of it are pretty obvious.

It's hard to put a figure on how much all this costs the taxpayers since we only know about the pilferers who get caught. And while we don't mean to be Pollyannas about this, the United States is far from a "kleptocracy" where *nothing*, not even the most routine business, gets done without a payoff. There are places like that in the world, and Transparency International's list of most- and least-corrupt countries is a good conversation starter, if controversial, at www.transparency.org. In 2006, Haiti came in at number one for corruption, with Guinea, Iraq, and Myanmar close behind. Finland, Iceland, and New Zealand were tied for the title of most honest nations on earth.[4] We've sat in focus groups with immigrants conducted by Public Agenda, the nonprofit organization we work for. For more than a few, the sense of relief at living under a government where they didn't have to bribe every cop every day was palpable.

Money for Nothing and Chicks for Free

Corrupt government employees who steal our money for themselves undoubtedly deserve their own ring in Dante's Inferno—and Dante was pretty clever with his punishments.[5] But frankly, lots of people outside the government find that ripping off taxpayers comes pretty easily. Tens of thousands of Americans suffered terribly from Hurricane Katrina, but for others, the disaster turned out to be a "let's-get-crazy" opportunity.

---

[4] Transparency International, "2006 Corruption Perceptions Index Reinforces Link Between Poverty and Corruption," November 6, 2006 (www.transparency.org/policy_research/surveys_indices/cpi/2006).

[5] Hoards of literature majors just raised their hands. Yes, Dante put crooked officials in Circle Eight. And it's pretty gross.

FEMA handed out debit cards so people displaced by the storm could get what they needed. And after a trauma like that, who's to say you don't need a Caribbean vacation? Or porn to get you through those lonely nights in the motel? And if the government is foolish enough not to check on whether you're collecting motel money and rental assistance at the same time, is that your fault? It's the sort of thing we expect criminals and people with me-first ethics to do whenever they get the chance. If the government's giving out "free" money, why not file a claim on the vacant lot next door. (Some people did this.)[6]

But lots of otherwise ordinary people seem to think it's OK to rip off large, anonymous institutions, and there's none bigger or more anonymous than the U.S. government. Kramer from *Seinfeld* is a man whose mind-set costs American taxpayers and consumers a lot of money. This, for example, was Kramer's stance on mail fraud.

KRAMER:	It's a write-off for them.
JERRY:	How is it a write-off?
KRAMER:	They just write it off.
JERRY:	Write it off what?
KRAMER:	Jerry, all these big companies, they write off everything.
JERRY:	You don't even know what a write-off is.
KRAMER:	Do you?
JERRY:	No. I don't.
KRAMER:	But *they* do. And *they're* the ones writing it off.

---

6 Government Accountability Office, "Hurricanes Katrina and Rita Disaster Relief: Improper and Potentially Fraudulent Individual Assistance Payments Estimated to Be Between $600 Million and $1.4 Billion," June 14, 2006.

Regrettably, the federal government is a frequent patsy for people who think this way. For example, getting a student loan is a rite of passage for many Americans, with nearly 10 million getting grants or loans in 2005–2006. And there are more than fourteen thousand students attending eligible institutions overseas, so no one at the Department of Education thought twice when three students put in for loans to attend the prestigious Y'Hica Institute for the Visual Arts in London.

They should have. It doesn't exist.

Oh, sure, Y'Hica had a Web site and a catalog, plus a letter from educational authorities in Great Britain certifying that it was an accredited institution. All of which are easily forged by anyone with a personal computer, in this case the staff of the Senate Governmental Affairs Committee, who were trying to find out how easy it would be to scam the government out of student loans. They got approval to seek $55,000 for three nonexistent students. To their credit, someone at Bank of America didn't like the looks of the application he or she received and turned it down. But both Nellie Mae and Sallie Mae swallowed the story whole. A top Education Department official sheepishly admitted the department "did not completely follow every step of the procedure."[7]

That's not the only time the government simply has failed to check on the massive amounts of applications coming in (and contracts going out). The Centers for Medicare and Medicaid Services, for example, admits that Medicare fraud is a problem. In fiscal 2004, the center believes Medicare paid out $900 million improperly for medical equipment, prosthetics, orthotics, and other supplies. The

---

[7] "Fake School Reveals Holes in Loan Program," Associated Press, January 21, 2003.

GAO says that while Medicare had set up a special system to check suppliers, it had flaws. For example, the "checkers" didn't bother to see if the suppliers were actually licensed and operating in good standing. Almost half of the $107 million spent on custom-made prosethics in Florida went to suppliers who didn't have their licenses checked, and forty-five of them were under investigation for fraud.

Category No. 2: Waste—or, Never Give a Sucker an Even Break

There are times when government behaves like a damn fool, either doing things that don't need to be done or doing potentially useful things in unbelievably wrongheaded ways. Mr. Matthews's $1,300 annual rice subsidies are like that. Crop subsidies have a purpose, or at least they used to have a purpose. During the Depression, farmers were going under in droves because they couldn't sell what they raised. Many ended up just walking away and letting the bank foreclose. (Remember the Okies of Steinbeck's *Grapes of Wrath*? Those guys were all too real.)

So the government came up with the idea of buying up surplus food and then paying farmers not to grow things they couldn't sell. That way the farmers didn't go broke and didn't have to lose their farms.

Extended Indefinitely

So what started as a response to a crisis has become an agricultural way of life. Eighty years later, agriculture is much more like big business, and by the mid-1990s much of the subsidy money was going to huge agricultural conglomerates.[8] Small farmers got subsidies too, but they also got

---

[8] "Federal Subsidies Turn Farms into Big Business," *Washington Post*, December 21, 2006.

a lot of federal regulations that kept them from switching crops easily. There was a good argument to be made that the system was actually hurting small farmers, not helping them. So the 1996 Freedom to Farm law introduced simplified direct payments, with the goal of weaning agriculture off subsidies. Farmers would get the money no matter what they farmed, or if they bothered to farm at all. And after seven years, the money would stop.

But a couple of years later, farm prices dropped, and under pressure from the farm lobby Congress extended the program indefinitely—and even increased the subsidies. And since the law promises payments even if nothing is grown, a suburban backyard that used to be a farm still qualifies. Which is why Donald Matthews gets his money. In parts of Texas that used to be big in rice growing, developers actually advertise their new subdivisions as being eligible for subsidies.

This, of course, is utterly boneheaded. But the Agriculture Department just shrugs and says it's implementing the law as written. And Congress, not wishing to tick off the farm lobby (which never wanted to lose subsidies anyway), is not inclined to change it.

That's usually the dirty little secret behind wasteful spending (and pork, which we'll talk about in a minute). It exists because somebody defends it—or at least defends the overall system, because he's afraid any changes might damage his interests.

I Can Get It for You Wholesale

Take Medicare (again). The inspector general of the Department of Health and Human Services concluded that Medicare pays double what the Veterans Affairs health care system pays for the same drugs and supplies. That's on average. In specific cases, Medicare pays a lot more: $8.68 per liter for saline, for example, compared with about $1.02 for the VA. You can get it cheaper at the local drugstore.[9]

Why? Because Medicare pays for supplies based on changes set in a 1987 law, which don't bear much relationship to current market value. When it comes to drugs, Medicare is obligated to buy "wholesale"—but the agency depends on the pharmaceutical companies to tell them what the wholesale prices are. And in the convoluted world of pharmaceutical pricing—which makes the fee structure for airline tickets look simple—there's no incentive for drug companies to make "wholesale" mean "the best price."

Lest you think that waste is all about big corporations with their hands out, let us set you straight. Sometimes government officials are plain just in over their heads.

Over at the FBI, they've been trying since 2001 to upgrade their computer systems. On TV's *Criminal Minds*, the chubby, punky computer genius Penelope Garcia can pull up complete dossiers on everyone, everywhere, without ever facing the "blue screen of death." And the team on *Without a Trace* seems to get all the computer info they need pronto, too. But there's a reason why that's called fiction. In the real world, FBI management has been famously technophobic. (In the 1990s, then-director Louis Freeh refused to use e-mail.) The old 1980s mainframe system was so bad that some agents avoided using it. The 9/11 Commission concluded the bureau's "woefully inadequate" technology and clunky policies on sharing information prevented agents from making sense of the scraps of seemingly unrelated intelligence that might have unraveled the 9/11 plot.

----

[9] Testimony of Inspector General Janet Rehnquist, Senate Committee on Appropriations, Subcommittee on Labor, Health and Human Services, and Education, June 12, 2002, available at oig.hhs.gov/testimony/docs/2002/020611fin.pdf.

"The FBI lacked the ability to know what it knew," the commission said.[10]

When the bureau finally started working to upgrade its computers, the FBI's leaders did what technologically naive and computer-phobic managers everywhere do: they were vague about their requirements and then kept tinkering with them; they set overly ambitious deadlines; and they neglected to make a plan to guide purchases and implementation. At least that's what the Justice Department inspector general concluded in a brutal 2005 report.[11]

$104 Million, $380 Million, $537 Million, and Counting

By 2005, the FBI realized that one key element of the project, the Virtual Case File, was so bug-ridden and behind schedule that it was unusable. That was $104 million thrown away right there, Director Robert Mueller III admitted to Congress. Overall, the GAO concluded that the bureau's upgrade costs had escalated from $380 million to $537 million. That's not counting the FBI's decision in 2006 to start over from scratch on an information-sharing system, contracting with Lockheed Martin at a cost of $305 million over six years.[12]

Unlike some, we're not arguing that government is inherently, laughably incompetent. We even suspect that at

---

[10] "FBI Organization and Priorities," chapter 3, National Commission on Terrorist Attacks Against the United States, available at www.911commission.gov.

[11] The Federal Bureau of Investigation's Management of the Trilogy Information Technology Modernization Project, Audit Report 05-07, Office of the Inspector General, U.S. Department of Justice, February 2005 (www.usdoj.gov/oig/reports/FBI/a0507/final.pdf).

[12] Harry Goldstein, "Who Killed the Virtual Case File," IEEE Spectrum, September 2005.

least some of you who work in the private sector are reading these stories and nodding, saying, "I can top that." There's a reason why *The Office* and *Dilbert* are so popular. As people who've spent our lives in the private and nonprofit sectors, we've got a few scary stories of our own.

And of course, the government certainly does not have a corner on fraud. Enron and WorldCom alone would prove that—and unfortunately, they're *not* alone.

But the difference is this: when Dunder-Miffin Paper Co. bungles things in *The Office*, a few employees and investors suffer. When the government bungles, we all suffer.

Category No. 3: Pork—or, Alexis de Tocqueville Has Left the Building

It's impossible to write an entire book about American politics without quoting Alexis de Tocqueville. In fact, in some states it may be illegal. We've come this far without mentioning the nineteenth-century French author who wrote one of the most famous books about the United States, but we can't hold it off any longer:

> The American Republic will endure until
> the day Congress discovers that it can bribe the
> public with the public's money.

That's the best definition of "pork" right there: the government bribing you with your own money. But one of the reasons we like this quote is that de Tocqueville was clearly wrong. That day has come and gone. In fact, it probably came and went not too long after he wrote this in the 1830s. At least according to some, the term *pork*, or *pork-barrel spending*, comes from the time when country politicians would gather at the general store, use a pork barrel for a desk, and pass out favors to the faithful.

Money Works Just as Well

It doesn't work quite that way now, thanks to civil service rules, sealed competitive bidding, and the fact that fewer people buy pork products by the barrel. But the idea is the same. To get elected to office—and stay there—politicians need to knit together a coalition adding up to 51 percent of the vote. Political science professors and editorial writers might like politicians to do that using reasoned debate, passionate oratory, and enlightened policy. But in the real world, money works just as well.

Let's face it, lots of voters see the federal government as a big sugar daddy. Why not? *Someone's* getting all those federal projects. And the benefits of those projects are real. If your community needs (or would like) a new bridge, and the federal government has money to pay for bridge building, why shouldn't you ask for it? Why shouldn't you expect your member of Congress to get it for you? And if he does, hasn't that earned him another term?

There are members of Congress, plenty of them, who view getting projects for their home states as a badge of honor. Consider Senator Robert C. Byrd (D-W.Va.), the longest-serving senator, author, zealous defender of the balance of powers—and, by general agreement, Congress's "King of Pork." "They don't know how much I enjoy it," says Byrd of that title, in much the same tone Robert Duvall uses to describe the smell of napalm in the morning. "I want to be West Virginia's billion dollar industry," Byrd said on another occasion, and no one doubts he is.[13]

Thanks to Senator Byrd, a slew of federal agencies that otherwise might have trouble finding Wheeling, West Virginia, on

---

[13] Michael Barone and Richard E. Cohen, *Almanac of American Politics, 2004* (National Journal Group, 2003).

a map now have major facilities there, from the FBI and Trea-sury Department to the Fish and Wildlife Service and NASA. And to express the state's appreciation—and just so voters don't forget their benefactor's name come election time—a lot of West Virginians see the senator's name nearly every day. There's the Robert C. Byrd Expressway, not to be confused with the Robert C. Byrd Freeway, *two* Robert C. Byrd Federal Buildings (call first if you're visiting to make sure you've got the right one), and the Robert C. Byrd Locks and Dam, along with Robert C. Byrd centers for science, health science, cancer research, tech-nology, "national technology transfer," and hospitality. And, of course, the Robert C. Byrd Federal Correctional Institution.[14]

In Senator Byrd's defense, if you grew up in a house with no electricity and no running water, and then pulled your-self up to represent one of the poorest states in the Union, you might feel obligated to bring home the bacon, too. The people of West Virginia seem happy with the deal—Byrd won his 2006 reelection campaign with 64 percent of the vote, carrying every county.

Does Anyone Really Need This?

So, defenders of pork ask, where's the harm? For example, NASA's Mission Control Center is in Houston for one reason: then vice president Lyndon Johnson was from Texas. NASA was going to build the center somewhere and Houston was as good a place as any. It hasn't done the space program any harm to have it there.

But there's a big difference between channeling needed projects to your home state, when and where you can, and

---

[14] Citizens Against Government Waste, "Words of Wisdom from the King of Pork," www.cagw.org/site/PageServer?pagename=news_byrd droppings, accessed December 3, 2006.

proposing projects that aren't needed, just to get more federal money into your state. To our way of thinking, this is one of the two big signs that pork has gotten out of control: when you hear about a project and can't imagine why *anyone* needs it or why it's the federal government's job to pay for it.

We haven't got anything against teapots or the people who collect them, but honestly, how many Americans really wanted their tax dollars spent on the Sparta Teapot Museum in Sparta, North Carolina? That got half a million dollars in the 2006 federal budget. How many of us are worried about increasing the number of wild turkeys (nope, they're not endangered)? The National Wild Turkey Federation got $234,000. And given the genuine challenges facing the country, how many of us really want government money spent on increasing winter sports? The Arctic Winter Games, a "circumpolar sport competition for northern and arctic athletes," got $500,000 from the defense budget, of all places.

Are "Earmarks" Really on Death's Door?

You've no doubt heard of "earmarks," the practice of a lawmaker putting special provisions into larger spending bills so they don't get the normal budget review. Earmarks have been around a long time, but their use exploded over the last dozen or so years. In that time, the number of earmarks tripled, costing taxpayers over $60 billion (*billion* with a *b*), and for many Americans, they became the preeminent example of Washington's penchant for pork. In the 2006 elections, the practice finally emerged as a political issue, and the new Democratic majorities in the House and the Senate promised to put some brakes on their use. There have been some changes (more about this in chapter 13), but there is also a fair bit of evidence that the practice (which serves members of Congress from both parties so well) is far, far from dead.

Indeed, earmarks have become an industry unto themselves, with lobbying firms that specialize in getting them springing up. Some of those lobbyists are out knocking on municipal doors, offering their services to communities whose requests for federal money have been turned down. Treasure Island, Florida, with a population of 7,514, had been rejected for federal money for bridge repairs. After spending $180,000 on its lobbyist, the town got $50 million for the bridge—when it had originally wanted only $15 million.[15]

So towns are spending local tax money in an attempt to get federal tax money to build unnecessary projects. Plus, local officials have to hire a specialist to get their congressman to listen to them—which is what he's elected to do, anyway.

That's the problem with pork.

IT'S JUST SO INFURIATING

So how much does all this actually add to the deficit? Most of the examples of waste, fraud, and pork we've talked about are counted in the millions of dollars, or even thousands of dollars, not billions. The *Congressional Pig Book Summary*, put together by people who are incensed about government waste and pork and not inclined to underestimate it, calculates that there was $29 billion in pork in the 2006 budget—but the deficit was $248 billion.[16] Obviously, it would be better to save that money rather than waste it, but hacking it out of the budget is not going to solve the country's financial problems.

---

[15] "Hiring Lobbyists for Federal Aid, Towns Learn That Money Talks," *New York Times,* July 2, 2006.

[16] Citizens Against Government Waste, *Congressional Pig Book Summary 2006,* available at www.cagw.org/site/PageServer?pagename=reports_pigbook2006.

The man who gave us the idea of "honest graft": Tammany Hall politician George Washington Plunkitt. *Credit: projects.ilt.columbia.edu*

Still, waste infuriates people, as much as or more than anything else the government does. Here we have to consult an expert in these matters, the late and less-than-honorable George Washington Plunkitt. Back in his day, a hundred years ago, Plunkitt was the very model of a Tammany Hall politician: paunchy, mustachioed, genial, and blatantly out for himself. Mustaches are out of style in Washington nowadays (although not paunches), but Plunkitt's spiritual descendants are still around. Plunkitt's important not because of jobs he held (he never rose above state senator) or bills he wrote. He's important because of a phrase he coined: "honest graft."

Graft, as you know, means politicians profiting from their connections with government. But Plunkitt prided himself on making a pile of money without breaking the law. "Dishonest graft" to him was actual bribery, embezzlement, or shaking people down, and he had contempt for it. "Honest graft" was another matter. Plunkitt's favorite scheme was to get advance information on where public works were being built, such as a new park, and buy up the land cheap before anyone else heard about it. Then he'd sell the land to the city for a nice profit. No laws were broken and the city wasn't paying any more than it would have paid the previous owner, so the taxpayers weren't being cheated, Plunkitt argued. He was even able to get offended when anyone suggested otherwise.

"I don't own a dishonest dollar," Plunkitt declared in 1905. "If my worst enemy was given the job of writing my epitaph when I'm gone, he couldn't do more than write: 'George W. Plunkitt. He Seen His Opportunities, and He Took 'Em.'"[17]

When Tammany finally got tossed out, our friend Plunkitt figured the problem was that the voters, foolishly, didn't get the difference between honest and dishonest graft.

"They saw that some Tammany men grew rich, and supposed they had been robbing the city treasury or levying blackmail on disorderly houses, or working in with the gamblers and lawbreakers," Plunkitt said. "As a matter of policy, if nothing else, why should the Tammany leaders go into such dirty business, when there is so much honest graft lying around?"

Interestingly, a hundred years later, many Americans worry that we haven't made much progress. When our organization, Public Agenda, conducted focus groups on the federal deficit and debt in 2006, federal finances weren't the top priority for most people when the groups started. It didn't

---

[17] "Honest Graft and Dishonest Graft: Very Plain Talks on Very Practical Politics," by Senator Plunkitt of Tammany Hall, recorded by William L. Riordon, 1905, available at www.panarchy.org/plunkitt/graft.1905.html.

take long for the participants to become concerned, but most were cynical and pessimistic about the government's ability to do anything about it.

Why? Because of a lifetime of hearing anecdotes about bungling, wasteful, inept government. It's no surprise that an ABC/Washington Post poll in 2002 found 61 percent said there was "a great deal" of waste in domestic programs and 43 percent said the same thing about military spending. Another ABC/Post poll in 2006 that asked Americans how much of their tax dollar was wasted found a median answer of 51 cents of every dollar—which by any objective estimate is way, way off the mark.

"I think to get robbed is a crime," one participant said in a New Jersey focus group. "To get robbed and lose trust, I think, is more of a crime, and that's what [politicians] are doing, they're taking our trust."

So Plunkitt *saw* it, but he didn't *get* it. People should be skeptical of their government; it's the only way democracy can work. But there also needs to be a minimum level of trust to get things done. Tammany Hall fell because the name "Tammany Hall" came to represent bloated, corrupt government. Eventually, people figured they couldn't trust Tammany to do anything right.

Given the budget mess we're in, the government should hang onto every dollar it can. But no one who's studied the federal budget believes that eliminating waste, fraud, and pork is going to balance the budget. Even if it did, that wouldn't do a thing to change the long-term demographic problems of Medicare and Social Security.

The real issue we face now is that we can't solve the budget crisis without trust—the public's trust that the government is facing the problem honestly, proposing real solutions, and that the inevitable sacrifices are fair to everyone. A lot of Americans may be willing to take a financial hit to ensure that Medicare is around to take care of their parents and, eventually, themselves. But why should they sacrifice if Medicare continues to pay more for drugs than you pay over the counter at Walgreen's? Why should people dig into their own pockets when Congress continues spending millions of dollars on bridges to nowhere just to up the reelection chances of the already powerful?

If a group of people get lost in the wilderness, and they pool their granola bars and Oreos so that no one goes hungry, that brings out their human dignity. But if the group is rationing snack food while one guy has a roast chicken hidden in his knapsack, that makes everybody else a sucker. And nobody puts up with being a sucker forever.

THE BEST AND THE BRIGHTEST

Once upon a time (1962, to be precise), the New York Mets were so hapless, so far down in the cellar, that manager Casey Stengel once wailed, "Can't anyone here play this game?"

Sometimes it's easy to feel that way about the federal government when you hear so much about waste and mismanagement. But it's also true that lots of federal programs run pretty well, and it's worth taking note of them. A good place to start is www.expectmore.gov, which is where the Office of Management and Budget posts its ratings of how well government agencies function. The OMB has a formal assessment process that

★★★★★★★★★★★★★★★★★★★★★★★★★★★★★★

asks twenty-five questions about whether federal agencies are achieving their goals and spending their money wisely. Based on that, the OMB rates programs as effective, moderately effective, adequate, ineffective, or "results not demonstrated."

As of mid-2007, the OMB said it had assessed 977 programs, covering almost all of the federal government. Of those, 75 percent were rated "adequate" or better, compared to 25 percent that were ineffective or had too little information to make a judgment. Only 17 percent got the highest rating, "effective," while 3 percent were "ineffective."

So who are those top performers? It's quite a mix. Some are comforting (the Secret Service protection programs get top marks, for example). Others are functions that frankly, we never would have thought of, but as soon as they're mentioned you realize they're really important to some folks. (Best example: the Coast Guard's domestic icebreakers. Somebody's got to keep shipping moving on the Great Lakes.) And, of course, there are frequently "effective" and "ineffective" programs in the same department.

Check out the entire list for yourself, but some themes that jumped out at us include:

★ **The statisticians:** One of the main products of the federal government is statistics about American life, and the government does that extremely well. The Census Bureau, Bureau of Labor Statistics, Bureau of Economic Analysis, Bureau of Justice Statistics, and the National Center for Education Statistics all got high ratings.

★ **The researchers:** A number of programs got high ratings at NASA and the National Science Foundation, including solar system exploration, astronomy and astrophysics, information technology, basic engineering, and polar research. The Federal

Aviation Administration's research on air safety is also on the list, as is the U.S. Geological Survey.

★ **Financial regulation:** Many of the programs that regulate banks and savings institutions got top ratings, including the Treasury Department's National Bank Supervision and Thrift Institution and Savings Association Supervision operations. The U.S. Mint and the Justice Department program that names bankruptcy trustees did well, too.

★ **People-to-people contact overseas:** The State Department gets good grades for its cultural exchanges, refugee assistance, and the consular services for Americans abroad. The Peace Corps does well, too.

SO MAYBE BILL GATES AND OPRAH COULD PAY OFF THE DEBT . . .

Nice try, but no way—not even if we vacuumed up away every single dollar they have.

It's not easy to grasp how much money we're talking about here, especially for those of us who get a sinking feeling every time we think about how many years it's going to take us to pay off our measly little home mortgages. Even the massive fortunes of Bill Gates and the Walton family (the Wal-Mart heirs) are

★★★★★★★★★★★★★★★★★★★★★★★★★★★★★★★★★

puny alongside the federal government's numbers. Take a look at the chart below for a sobering little lesson in the numbers we're up against.

The federal government's red ink for 2006: $248 billion

The combined fortunes of the ten richest people in America: $232.9 billion[18]

Bill Gates, Microsoft, Medina, WA: $53 billion

Warren Buffett, Berkshire Hathaway, Omaha, NE: $46 billion

Sheldon Adelson, hotel and casino entrepreneur, Las Vegas, NV: $20.5 billion

Lawrence Ellison, Oracle, Redwood City, CA: $19.5 billion

Paul Allen, Microsoft, Vulcan, Seattle, WA: $16 billion

Jim C. Walton, Wal-Mart, Bentonville, AR: $15.7 billion

S. Robson Walton, Wal-Mart, Bentonville, AR: $15.6 billion

Christy Walton and Family, Wal-Mart, Jackson, WY: $15.6 billion

Alice Walton, Wal-Mart, Fort Worth, TX: $15.5 billion

Michael Dell, Dell, Austin, TX: $15.5 billion

Total outstanding debt of the U.S. government: roughly $9 trillion

Combined wealth of the Forbes 400 wealthiest Americans: $1.25 trillion

And as rich and successful as they are, people like Oprah and Donald Trump couldn't even keep up with the federal government's interest payments. Oprah's fortune is estimated at about $1.5 billion. Donald Trump is a little richer at $2.9 billion. The federal government? In 2006, it spent more than $226 billion just on interest.

---

[18] "The 400 Richest People in America," 2006 edition. *Forbes* Special Issue, October 9, 2006.

The Perpetual Disaster Relief Dilemma

The mistakes FEMA (the Federal Emergency Management Agency) made responding to Hurricane Katrina will likely be dissected for decades. But even when the disaster is less horrific and government disarray is less infamous, FEMA's history illustrates how complicated it can be to root out fraud in federal spending. When disaster strikes, we in effect ask government to do two somewhat different things. On the one hand, we want government to get money to people quickly so they can rebuild their homes and put their lives back in order. On the other, we want

Hurricane Katrina brought out the best in some and the worst in others. *Credit: Federal Emergency Management Agency*

it only to give money to people who have actually suffered—not to crooks angling to get in on the getting while the getting is good.

Unfortunately, disaster relief in the United States (and probably elsewhere) often follows a familiar pattern. Right after a flood or earthquake or hurricane, there is a push to help people generously and quickly. A few months later, reports that millions of dollars have gone astray start to pour in. Almost inevitably it seems, the government has given money to people who weren't really there, who collected two or three payments for the same thing, or who damaged their own property in order to get something new—people who worked the system.

When Hurricanes Ivan, Charley, Frances, and Jeanne crashed their way through Florida in 2004, the results were devastating. A helicopter survey conducted by the *Miami Herald* showed "structural damage to nearly every beachfront home along a 30-mile stretch of coast . . . crushed mobile homes, shredded roofs, overturned cars . . . swamped boats, washed out bridges, flipped private airplanes . . . more than 2.5 million customers without power, in some cases for weeks."[19]

Not surprisingly, President Bush promised aid and Congress quickly approved more than $2 billion in relief. Democrats in the state called the money "inadequate" and held a press conference to protest.[20]

Yet just weeks later, the headlines had changed. Although Hurricane Frances "largely missed" Miami-Dade County, according to the local *Fort Lauderdale Sun-Sentinal,* over ten thousand residents successfully put in claims for storm damage, totaling over $25 million.[21] As the newspaper reported, FEMA paid for "new cars, wardrobes and thousands of televisions and appliances—even though the storm brought wind and rain no worse than a thunderstorm.[22] Some local residents were seen "pouring water on furniture and throwing rocks at cars to make it look like [hurricane] damage."[23] Some referred to FEMA payments as "Christmas money" or "free money."

The problem, as one FEMA official described it, is the need to walk "a fine line between speedy service to those who need it and

---

[19] Mercury News Wires Services, "Fourth Storm Devastates Battered Florida Coast," *San Jose Mercury News,* September 27, 2004.

[20] Frank Davies, "Storm Relief to Benefit Florida," *Miami Herald,* October 6, 2004.

[21] Sally Kestin, Megan O'Matz, and Luis F. Perez, "Training for FEMA Inspectors Often Brief," *Sun-Sentinal,* October 31, 2004.

[22] Sally Kestin Megan O'Matz, "Suspicions Don't Slow Miami-Dade Storm Relief," *Sun-Sentinal,* December 5, 2004.

[23] Kestin, O'Matz, and Perez, "Training for FEMA Inspectors Often Brief."

ensuring that taxpayer dollars are not misused."[24] But the line seems continually blurred. Similar problems occurred after September 11, after Hurricanes Katrina and Rita in 2005, and elsewhere. To Maine senator Susan Collins, speaking during congressional hearings on the problem, "This pay-first, ask questions later approach has been an invitation to the unscrupulous."[25] But no one seems to have come up with good ways to verify identities, confirm Social Security numbers and addresses, and ensure that losses really are a result of the disaster at hand right when the disaster is still at hand.

And sometimes attempts to save money and be more helpful just make things worse. Right after Hurricanes Katrina and Rita, FEMA gave out $2,000 debit cards to storm victims to avoid spending time issuing and mailing out thousands of government checks. Among the charges government auditors later found on the debit cards: $450 for a tattoo, $1,300 for a handgun, $150 for adult erotica, $1,000 for a bail bond, and $1,100 for a diamond ring.[26]

Still, if it's any comfort to the poor, besieged taxpayer, the government is not alone. Charitable groups like the Red Cross have their share of fraud, too. According to a 2005 Red Cross press release, "fraud is a part of life following natural disasters." In the release, the group reported that it was centralizing its record keeping to guard against people who claim relief from different Red Cross sites or "'shop' for higher levels of assistance from different Red Cross service centers."[27]

---

[24] Sandy Coachman, federal coordinating officer for FEMA, quoted in Federal Emergency Management Agency (FEMA), "Disaster Recovery Officials Issue Alert for Fraudulent Claims," Department of Homeland Security Documents, November 17, 2005.

[25] Megan O'Matz, "FEMA Phone Fraud Detailed," *Sun-Sentinel*, February 14, 2006.

[26] Ibid.

[27] American Red Cross, "American Red Cross Committed to Fighting Fraud," PR Newswire, September 29, 2005.

BOONDOGGLING FOR FUN AND PROFIT

Once upon a time, a boondoggle was a way of tying a lanyard or neck-
erchief. Despite what you might have heard in the movie *Napoleon
Dynamite,* boondoggles were never "a must have for this season's
fashion." But in 1935, the New York City Board of Aldermen was
somewhat nonplussed to find out that the government was paying
people to make them. In fact, some $3 million of city relief funds
were being spent on teaching unemployed people to dance and, as
the *New York Times* put it, "otherwise amuse themselves."[28]

This was in the midst of the Great Depression, with one in
five workers unemployed. It's not so much that the government
felt the jobless needed to jitterbug, but President Roosevelt's
New Deal strategy was that if there were no private-sector jobs
for these people, the government had better come up with jobs
for them, and fast. So the government set up whole agencies
just to put people to work. The Works Progress Administration employed millions of people building schools, post offices,
parks, and town halls across the country. The Civilian Conserva-
tion Corps planted trees and tended national parks. The Federal
Writers' Project and Theatre Project funded the arts, including
rising young talents like James Agee and Orson Welles.

But it fell to a crafts instructor named Robert Marshall to
face the music on boondoggles.

"I spend a good deal of time explaining it," he said ("some-

---

[28] "$3,187,000 Relief Is Spent to Teach Jobless to Play; $19,658,512
Voted for April," *New York Times,* April 4, 1935.

what sadly," the *Times* noted). It could have been worse. He appeared right after Myra Wilcoxon, who was teaching people "eurythmic dancing." The board members asked her to demonstrate a few moves, but she refused.

So the boondoggle, once a perfectly innocent technique taught to Boy Scouts, became a term for anything big, useless, and funded with government money.

When you think of a boondoggle, you may be thinking of something like this:

Way behind schedule and far over budget, the "Big Dig" highway project in Boston cost $12 billion more than originally projected. *Credit: Massachusetts Turnpike Authority*

It's Boston's "Big Dig," also known as the Central Artery/ Tunnel, a massive construction project to change Interstate 93 from an elevated roadway to a 3.5-mile tunnel under the city. At $14.6 billion, it's the most expensive highway project in U.S. history. And it came in a mere $12.1 billion over the original projections. But that's not counting $100 million the state

★★★★★★★★★★★★★★★★★★★★★★★★★★★★★★★★★★★

attorney general is trying to get contractors to give back because of shoddy workmanship. In fact, a few months after the tunnel was opened, it had to be closed again because a piece of concrete fell off and killed someone.

A boondoggle is pretty much in the eye of the beholder. Some people say farm subsidies, the international space station, and the Pentagon's missile defense plan are boondoggles, but they've all got their defenders. In this case the Big Dig is, in fact, open for business, and it's easier to get across Boston than it used to be. To us, that means it's not useless and therefore, technically, not a boondoggle. It's merely badly built and vastly over budget.

But it's a close call.

CHAPTER 11

The Liberals and the Conservatives

Blessed are the young, for they shall inherit the national debt.

—Herbert Hoover (1874–1964), U.S. president

Given the threat to the nation's finances, you might think the president and Congress—the Republicans and the Democrats, the liberals and the conservatives—would be up there wheeling and dealing, looking for areas of agreement and compromise. Research conducted by the organization we work for, Public Agenda, shows that most Americans believe that we will have to compromise on this issue.[1] Compromise has an honorable tradition in American government, and politicians are generally pretty good at horse trading.

Logically, after all, the main options for closing the budget gap are tax increases or spending cuts, so the debate in Washington ought to be about how much of each to apply,

---

[1] Public Agenda, *It's Time to Pay Our Bills: Americans' Perspectives on the National Debt & How Leaders Can Use the Public's Ideas to Address the Country's Long-Term Budget Challenges,* April 2006 (www.publicagenda.org/research/pdfs/facingup_leaders_report.pdf).

right? There are few issues more amenable to a practical, pragmatic approach than the federal budget. You'd think.

But we haven't heard much talk of compromise in Washington in recent years. Instead, what most of us hear most of the time is a battle of the sound bites—usually over-heated, over-the-top sound bites.

President Bush has signed major tax cuts into law during his presidency, and for him, the idea of increasing taxes would be "disastrous for business, disastrous for families and disastrous for this economy."[2] But disaster appears to be in the eye of the beholder. For *New York Times* columnist and longtime Bush critic Bob Herbert, it's the president who's created the economic disaster: "President Bush and his clueless team of economic advisers" are "the ferociously irresponsible crowd that has turned its back on simple arithmetic and thinks the answer to every economic question is a gigantic tax cut for the rich."[3]

Whew! President Bush and Bob Herbert offer up some stirring phrases there, and we give both of these gentlemen credit for being sincere in their beliefs. But occasionally you have to ask yourself just how far this kind of rhetorical fire-breathing gets us.

Maybe it's the political equivalent of the old "hellfire and damnation" sermons—designed to scare the congregation into repenting. Sometimes it reminds us of Bill Murray forecasting doom in *Ghostbusters:* "This city is headed for a disaster of biblical proportions . . . Human sacrifice! Dogs and cats living together! Mass hysteria!" Of course, in the

[2] "President Bush Discusses Strong and Growing U.S. Economy," speech to the American Council of Engineering Companies, May 3, 2006, available at www.whitehouse.gov/news/releases/2006/05/20060503-4.html.

[3] Bob Herbert, "No Work, No Homes," *New York Times,* August 14, 2003.

movie, New York faced destruction at the hands of an evil Sumerian god shaped like a gigantic Stay-Puft Marshmallow Man. Definitely a crisis worth getting excited about.

Politicians and commentators insulting each other makes for a good read and seems to improve TV ratings, but we suspect a lot of Americans are getting tired of it. We don't think it's a coincidence that polls show Congress, the presidency, and the press getting some of their lowest performance grades in recent history. Outside the Beltway, most Americans seem ready to move on, and so are we. The country's financial mess doesn't have to lead to disaster if the country stops screaming and starts working on the problem now.

Still, it's important to remember that the two sides didn't get mad at each other over nothing. For all the hyperventilating among the partisans, both liberals and conservatives bring important ideas about government and the economy to the table. So we've done for you what congressional staff members often do for their bosses. We've created a quick recap of the main points liberals and conservatives make about matters budgetary (known as "talking points" in Beltway-speak).

If you've been getting your news mostly from your own politically comforting source (be it Salon, Fox News, Ann Coulter, or *The Colbert Report*) here's your chance to refresh yourself on the other side. And even if you plan to steer a course right down the political middle, it's not a bad idea to keep the shorelines in sight.

So here are "the rules" as far as most conservatives see them:

1. Individuals and families, local governments, and the marketplace almost always do a better job of solving problems than the federal government.

2. The best federal government would be a much smaller one, covering national defense, homeland security, and a few other items that local governments or the private sector simply can't handle.

3. When the federal government tries to "help" in areas like retirement, health care, prescription drugs, education, economic development, and so on, it wastes money and mangles the job.

4. Waste, fraud, abuse, and mismanagement are rampant in the federal government today. When a government official is allowed to spend other people's money, this result is almost inevitable.

5. Social Security and Medicare take money out of workers' hands and give them unstable, unfair federal programs in return.

6. Nearly all Americans would be better off if Social Security and Medicare were redesigned as private savings programs. That would capitalize on Americans' sense of individual responsibility and judgment.

7. Low taxes lead to more investment and entrepreneurship—that helps the economy grow.

8. When people and businesses earn more, they return more to the government in taxes.

9. Keeping the economy growing is the way to improve Americans' standard of living.

10. Big government and high taxes interfere with the marketplace. When the free enterprise system is left to operate without interference, it produces the prosperity that benefits us all.

And for the liberals:

1. Government has a responsibility to try to improve people's lives and address problems like inequality and poverty.

2. Americans need more government help, not less. Number one example: the government should help the nearly 45 million Americans without health insurance.

3. It's fashionable to complain about "big government," but Americans need and want programs like Social Security, Medicare, Head Start, food stamps, the Centers for Disease Control, and many others. They do things that need to be done and do them well.

4. Social Security and Medicare are two of the great advances of the last century. Before them, millions of older Americans lived in poverty with inadequate health care.

5. Conservative plans to "privatize" Social Security and Medicare are unworkable. Wealthy, well-educated people will benefit. Poor people, not-so-savvy investors— even people who just have a run of bad luck—will be much worse off.

6. Taxes are the way we pool our resources so government can address society's problems.

7. But taxes are not fair now. The recent tax cuts have mostly benefited the very wealthiest Americans. These tax cuts are the major cause of our current budget problems.

8. Wealthy Americans who prosper from living and working in this country should be willing to pay more. They should pay higher taxes so less-fortunate Americans can enjoy some of the advantages they have had.

9. What threatens the economy isn't taxation, but leaving millions of Americans in underfunded schools, in poor neighborhoods, and without decent medical care. If those people can't find decent work and get decent services, they'll never contribute what they could to society.

10. Competition and the free market are good things, but they can't solve every problem, and they don't always work for everyone.

A LITTLE OF BOTH, PLEASE

If you read these over and find yourself agreeing with some ideas from both lists, you're not alone. Moderates and middle-of-the-roaders don't get much respect these days, but mixing and matching is totally allowed in this book. In fact, we're convinced that a little mixing and matching is the only way the country is going to work its way back to financial sanity.

What you're not allowed to do is adopt a conservative "keep taxes as low as possible" stance, while blithely endorsing liberal ideas to spend money on good works. Sadly, this is a compromise that a lot of D.C. types seem willing to live with, which is unfortunate, because it's the one compromise that can't possibly work. That's when the numbers just don't add up.

CHAPTER 12

Politics, as Usual

Politics is supposed to be the second oldest profession. I have come to realize that it bears a very close resemblance to the first.

—*President Ronald Reagan (1911–2004)*

Let's just suppose that Americans across the country began thinking that we need to get serious about the nation's finances. And let's say that large majorities started saying things like "This is serious. We're going to have to compromise and do some things we don't really like, so let's get on with it." Would anything actually happen in Washington, D.C.?

Being the optimists that we are, we believe there are enough reasonably sane and decent people in Washington who are ready to work hard to get the country's finances back on track. But for that to happen, normal-type people—those of us who don't run for office and don't live and breathe politics on a daily basis—need to understand the reality of the situation. And as citizens and voters, we need to start doing some things a little differently ourselves. Like any other issue the country faces, there are politics involved here. We're optimists, but we're not lunatic enough to think that politics doesn't matter.

Boiling the problem down to its essentials, what this means is that the people now in office—and those running for office—have to believe that they would be better off addressing the country's deficit and debt problems responsibly rather than avoiding or spinning them. That's our challenge as citizens.

And we believe this can be done. But there are a lot of powerful forces tugging on an officeholder every day, and very few of them are pulling in the direction of fiscal responsibility. Many of these factors are actually making things worse. So you need to know what you're up against.

President Harry Truman, first a politician, then a statesman." *Credit: Library of Congress*

A politician is a man who understands government. A statesman is a politician who's been dead for 15 years.

That line comes to us courtesy of Harry Truman, who has long since been promoted to statesman status. But we wouldn't remember that quote, or anything much about Truman, if the voters of Missouri had decided he should remain a menswear-store owner who liked to talk politics. Truman understood that the first job of any politician, from president to dogcatcher, is to get elected. And getting elected often means doing things that only make sense if you're running for office. For example, Truman, on his way up, played ball with Missouri's Prendergast machine, as dirty a bunch of operators as ever stuffed a ballot box.

In the novel *Primary Colors,* Governor Jack Stanton (a thinly disguised version of Bill Clinton) is challenged by one of his aides on the hardball maneuvers Stanton's pulled trying to win the presidential nomination.

"We live an eternity of false smiles—and why? Because it's the price you pay to lead," Stanton said. "You don't think Abraham Lincoln was a whore before he was a president? He had to tell his little stories and smile his back-country grin. He did it all just so he'd get the opportunity, one day, to stand in front of the nation and appeal to 'the better angels of our nature.' That's where the bullshit stops. And that's what this is all about. The opportunity to do that, to make the most of it, to do it the right way—because you know as well as I do there are plenty of people in this game who never think about the folks, much less their 'better angels.' They just want to win . . . Is there anyone out there with a chance to actually win this election who'd even think about the folks I care about?"

Now, you may think that's the most self-serving rationalization you've ever heard. But the thing is, *politicians* believe that, and you can't understand them without understanding that point of view.

Every politician tells himself (or herself), in his introspective moments, that he does the more ridiculous things he does so he can win the election and start doing good. In politics, getting elected is the bottom line. If you can't get into office and stay in office, you can't do anything for your constituents or your state.

Rightly or wrongly, most members of Congress think the candidate from the other party who would take their seat if they lose has awful ideas about what government should and should not do. So, from their perspective, either they do what it takes to get reelected or someone worse will take their place. Same goes for people running for president. You probably can't go through what you have to go through in a presidential election campaign unless you are utterly convinced that the other candidates would do a much worse job than you would do. And the need to keep doing whatever it takes (within reason, of course) to get elected doesn't stop after you win the first time.

ALWAYS RUNNING, RUNNING, RUNNING

Now we all know (we do all know this, don't we?) that senators have to run for office every six years, but members of the House have to run every two years. Senators have a little more time to catch their breath between elections, but they also generally need to raise more money to campaign since they have to run statewide. House members face a fairly exhausting future if they're in a competitive district. Suppose you had to reapply for your own job every two years—with all the interviewing and wearing your best clothes and having butterflies in your stomach every time the phone rings. You know there's another applicant who wants your job very badly and is really making nice with your boss. Even worse, there seem to be plenty of people around who also seem to think this person would be better at your job than you are.

Essentially, House members start the next campaign almost as soon as the last one is over. At the very least, they have to start fund-raising. It takes a lot of time and effort to raise the kind of money it takes to run for Congress these days. The average House candidate spent $646,485 in the 2006 campaign, and the average Senate candidate spent $3.3 million. And that's the *average*—the candidates in the most expensive House race (Florida's 13th District) spent $11.1 million all told.[1]

And when you raise money, that means going to people who have money—and a particular interest in who serves in Congress. Big donors get the most attention here. The finance, insurance, and real estate industries gave $251 million to candidates in 2006, and ideological/single issue groups gave another $182 million. But smaller donors are playing a larger role. Many of the key donors are also networks of smaller donors, such as EMILY'S List, which funds promising women candidates (but mostly Democratic ones). Still, less than 1 percent of the American public gives money to candidates. And the one thing that nearly all those donors have, big or small, is expectations. At a minimum, they expect their representative to talk a lot about their concerns. And if a representative doesn't support their views, they're not likely to give to him or her again. Please understand—we're not saying this is automatically bad. In fact, the Supreme Court considers this a form of free speech. But it is, shall we say, constraining.

We're not suggesting that you need to get all teary-eyed over the rough lot of members of Congress. They get plenty of perks, such as eating in the congressional dining hall and getting to use the swimming pool and special elevators and all. They're even called the "honorable" so-and-so, whether they're actually honorable or not.

---

[1] Center for Responsive Politics, 2006 Campaign Spending Overview, www.opensecrets.org/overview/stats.asp, accessed June 10, 2007.

CAMPAIGNS BUT NO COMPETITION

What's more, a lot of House races just aren't that competitive. Sometimes the representatives have done a good job year after year and won the loyal support of voters; sometimes they've benefited from redistricting that gives them the best shot possible. Both parties finagle with redistricting designed to ensure "safe" seats for their incumbents, but the practice has grown stronger in recent years.[2]

Most of the time, congressional seats turn over because of death, retirement, or scandal—not because voters got fed up with the actual positions of the incumbent. Consider the 2006 congressional elections, by any standard a major shift in political fortunes. Democrats took control of the House of Representatives after twelve years in the minority, riding a wave of public discontent over the war in Iraq, congressional scandals, and the Bush administration. Yet in that shift, only 30 House seats out of the 435 up for election actually changed from Republican to Democrat.[3] And while an unusual number of incumbents were running scared that year, the vast majority coasted to reelection. Before the 2006 vote, the respected and nonpartisan Cook Political Report listed 347 House seats as "solid," meaning "not competitive at all."[4] The party that has those seats will likely still have them after the 2008 election as well.

---

[2] Common Cause is a major critic of the way redistricting is handled. You can find out more at their Web site: www.commoncause.org (www.commoncause.org/site/pp.asp?c=dkLNK1MQIwG&b=196481).

[3] Cook Political Report, "2006 House Election: Seats That Switched Parties," December 13, 2006, available at www.cookpolitical.com/races/report_pdfs/2006_house_turnover.pdf.

[4] Cook Political Report, "2006 Competitive House Race Chart," November 8, 2006, available at www.cookpolitical.com/races/report_pdfs/2006_house_comp_nov8.pdf.

EVERYTHING YOU DID IN GRADE SCHOOL

Still, it's worth it to keep in mind that senators and representatives are human beings, and a lot of them went into politics hoping to do some good. Running for office means asking everyone you know for money, exposing yourself and everything you have ever said or done since grade school to public scrutiny, and going to diners at the crack of dawn, interrupting people eating breakfast and urging them to vote for you. So they do have their own tough road to travel.

You might ask why a politician in a "safe seat" would worry so much about being reelected. They should be more independent, you say, more willing to take on tough issues, not less. And sometimes they are. But there are plenty of examples of politicians who thought they were safer than they really were—until they found out differently at the polls. A safe seat is not a blank check. In fact, there's a school of thought that says being in a safe seat makes it harder to compromise, not easier. If you're a member of Congress representing a safe district, a substantial number of your constituents are probably committed party members—the kind of folks who get involved, blog, give money, and get really, really irritated when "their representative" strays from the party line.

AND THE PUBLIC'S NOT MUCH HELP, EITHER

But on this issue, the hurdles are even higher. It's not just the party faithful and political opponents who can do you in. On this issue, the great American public is not much help, either. Here's the sad truth. For someone who needs to attract the attention and gain the support of the voting public, talking about the nation's finances and suggesting realistic ideas for improving them is just not an attractive option. Consider this:

TO MOST AMERICANS, OTHER ISSUES MATTER MORE

The deficit and the debt just don't make it into the winner's circle when surveys ask Americans what problems they're worried about. Most people name terrorism or the war in Iraq or the economy as the top issue in their mind. Even when pollsters read people a long list of possibilities and ask people about each of them, nearly every issue you can think of ranks higher than reducing the federal deficit—education, jobs, economic growth, Social Security, Medicare, crime, health care, energy, environment.[5]

All these issues are important, and we certainly expect the candidates we vote for to talk about them. Plus, from the campaign manager's point of view, showing your candidate working on problems like these can make such a nice political ad. There he or she is (with inspiring music in the background) reading to grade-school kids, congratulating new college grads, chatting with senior citizens, chewing the fat with factory workers or farmers, walking beside a sparkling river. It's so much easier than producing an ad in which your candidate says, "I plan to cut spending on popular programs and raise taxes in order to cut the deficit."

Even worse for political types, current polling shows that most Americans say that it's worth *increasing* the deficit in order to spend more on terrorism, the military, schools and colleges, and stimulating the economy.[6] With that in mind, the men and women who need happy voters to keep their jobs are often inspired to do just that—they increase

[5] Princeton Survey Research Associates for Pew Research Center, January 4–8, 2006, available at www.publicagenda.org. See "Federal Budget: People's Chief Concerns," under "Issues."

[6] Ipsos-Reid for the Committee for Education Funding, February 1–2, 2002, available at www.publicagenda.org. See "Federal Budget: Bills and Proposals," under "Issues."

the deficit to spend more on the things people like. Even the braver politicians, those who might be willing to sacrifice votes to take this on, may stop and wonder whether people will feel much gratitude if they succeed.

PUBLIC OPINION LAND MINES EVERYWHERE YOU LOOK

It's not just that reducing the deficit is not a top priority for most Americans. It's that public opinion is just all over the place on this issue, and that makes political types nervous. No matter what a politician does, he or she runs the risk of making a lot of potential voters unhappy. For example, Congress passed some pretty hefty tax cuts in 2001 and 2003, but in 2006, half of all Americans (51 percent) still said that cutting income taxes for the middle class should be a top priority for Congress and the president.[7] Just one in four would support raising taxes to reduce the budget deficit; seven in ten say they oppose this approach.[8] Most politicians just aren't too eager to rile these folks.

Meanwhile, 80 percent of Americans say it's the government's responsibility to "provide a decent standard of living for the elderly."[9] Numbers like that strike fear into the heart of any politician who even thinks about proposing Social

---

[7] Pew Research Center for the People and the Press, "January 2006 News Interest Index, Final Topline," January 4–8, 2006 (people-press .org/reports/questionnaires/268.pdf).

[8] Pew Research Center for the People and the Press, "October 2005 News Interest index, Final Topline," October 6–10, 2005 (people-press .org/reports/print.php3?pageID=1009).

[9] CBS News/New York Times Poll, June 10–15, 2005: "On the whole, do you think it should or should not be the government's responsibility to provide a decent standard of living for the elderly?" Should: 80 percent; should not: 16 percent; unsure: 4 percent. Available at www .pollingreport.com/social.htm.

Security or Medicare cuts. Then, nearly six in ten say they oppose reducing defense and military spending to cut the deficit.[10]

Let's see. Can't raise taxes. Can't touch Social Security and Medicare. Can't touch defense. There's not all that much left in there to cut. (And if you think you can stop the red ink by eliminating "waste, fraud, and abuse," you must have skipped chapter 10. Go back and read it.)

JUST GIVE PEOPLE A MOMENT TO THINK

Let's not let the pollsters have the last word on this, however. Opinion polls often capture people's top-of-the-head responses—the very first thing that flies out of a person's mouth. Give someone a little time to think it over, even a few minutes, and you might get a different answer. We work for an organization that conducts opinion research, so we've seen this often. Research conducted by our organization, Public Agenda, and a California-based research group, Viewpoint Learning, suggests that many Americans are more realistic about the country's financial problems once they have the chance to think them over even for a relatively short period of time.

Viewpoint Learning brought typical citizens together for daylong discussions on what should be done to address the money crunch the country faces with the aging of the boomers. Most participants showed a lot of openness and flexibility about different ways to solve the problem. At the end of the day, most believed that the country would need

---

[10] Pew Research Center, March 2005: "Would you favor or oppose lowering defense and military spending as a way to reduce the budget deficit?" 31 percent favor; 60 percent oppose; 3 percent unsure. Available at www.pollingreport.com/budget.htm

to raise taxes and cut spending, and most were prepared to back some of both.[11]

BETWEEN THE SWORD AND THE WALL

But the problem in politics is that most Americans aren't spending a lot of time talking or thinking about the country's yearly deficits, the growing debt, or the costs government will face from the aging boomers. Most Americans aren't keen on raising taxes. They like Social Security and Medicare the way they are (and maybe even a little better, please). They don't want to cut federal spending on defense or education or any manner of good things.

That puts a responsible candidate or lawmaker who wants to make sensible proposals in a bind. It's the political equivalent of being between a rock and a hard place. In Spanish, the phrase is being between "the sword and the wall," which conjures up visions of Zorro slashing a "Z" into some chubby guy's pants. It's a more menacing image, but luckily, not that many members of Congress speak Spanish.

FIBS, LIES, AND VIDEOTAPE

Some politicians thrive on campaigning, drawing energy from the long hours and give-and-take. Others, more privately, view elections as something to be suffered through on the way to office. What virtually no one likes is "negative campaigning" that twists and distorts a candidate's record and positions. If you've watched the news during an election season, you know the ads we're talking about. They might

---

[11] Steven A. Rosell, Isabella Furth, and Heidi Gantwerk, *Americans Deliberate Our Nation's Finances and Future: It's Not about Taxes—It's about Trust* (Viewpoint Learning, 2006).

feature some gape-mouthed, haggard photo of a candidate, a voice-over announcer who also does horror-movie trailers and the clear message that electing this candidate would be the moral equivalent of having your state overrun by giant mutant locusts. Then, at the end, there's an incongruously chipper disclaimer from The Other Guy: "I'm Millard Fillmore and I approved this message!" Negative campaigning is a big, big complication in the politics of the budget. It's not really a mystery why.

Since most Americans say they don't want their taxes raised, accusing your opponent of supporting higher taxes gets you points. And since most Americans say they don't want cuts in popular areas like Social Security, Medicare, defense, education, and so on, accusing your opponent of cutting those gets you points. And since negative campaigning plays so fast and loose with the truth, even modest attempts to compromise on taxes and spending can come back to haunt you during your next campaign.

FACTCHECK AND "THE WHOPPERS OF 2004"

In 2004, President Bush and Senator John Kerry were both perpetrators and victims of this kind of grubby campaigning. The University of Pennsylvania's Annenberg Public Policy Center runs a Web site called FactCheck.org that keeps tabs on the tall tales candidates tell when they're running for office. If you have a strong stomach, you might want to take a walk down memory lane in FactCheck's "Whoppers of 2004" section. Be warned, though—if you're some crazed fanatic who thinks candidates should actually talk about the pros and cons of their ideas and proposals when they run for office, it can be an exasperating trip. So here's just a small sampling from FactCheck's review of the 2004 presidential campaign.

This is President Bush in Florida on March 20, 2004.

PRESIDENT BUSH:	Senator Kerry is one of the main opponents of tax relief in the United States Congress. However, when tax increases are proposed, it's a lot easier to get a "yes" vote out of him. Over the years, he's voted over 350 times for higher taxes on the American people—
AUDIENCE:	Booo!

The allegation that Senator Kerry had voted for higher taxes 350 times was repeated over and over by Bush campaign operatives. But when FactCheck asked for a list, it was "padded with scores of votes Kerry cast against tax decreases (which would leave taxes unchanged, not higher), votes to reduce the size of proposed tax cuts (which would leave taxes *lower*, though not as low as proposed), and 'votes for watered-down, Democrat "tax cut" substitutes' (which often proposed to distribute the benefits of tax cuts farther down the income scale than Republican proposals). Thus, the Bush campaign counted some votes for tax *cuts* as votes for 'higher taxes.'"

This distortion of Kerry's record was so egregious and so widely panned in the press that even staunch Bush supporters stopped repeating "350 times" after a while, but the damage to Kerry was probably already done. And for worried politicians in tax-averse districts, the message is crystal clear. It's not enough to vote for tax cuts; you have to vote for the biggest tax cut possible or you could find yourself in Senator Kerry's shoes.[12]

---

[12] FactCheck.org, "Bush Accuses Kerry of 350 Votes for 'Higher Taxes': Higher Than What?" March 23, 2004, updated March 24, 2004 (www .factcheck.org/article159.html).

Meanwhile, the Kerry campaign produced its own slash-and-burn version of the truth—an ad accusing President Bush of planning to cut benefits for people on Social Security. Again, from FactCheck:

ANNOUNCER: They were hoping to keep it a secret, but we just learned that George Bush and the Republicans are planning to privatize Social Security after the election. Bush and the Republicans have already put Social Security at risk with a record deficit of over $400 billion. Now Bush and the Republicans have a plan to privatize Social Security that cuts benefits by 30 to 45 percent.

Bush and the Republicans, a plan to cut Social Security benefits.

Give us a break. There was nothing secret about President Bush's interest in setting up private accounts as part of Social Security; he had been talking about it for years. Please. And no matter what you think of the president's idea, the fact is that he repeatedly emphasized that benefits for people on Social Security now would not be reduced in any way.[13] Yet that is exactly what the Kerry ad implies.

You see how it goes. This is why elected officials get out of Dodge anytime someone raises the prospect of curbing Social Security or Medicare spending. Just look how it could be presented to voters when the next election comes up.

---

[13] FactCheck.org, "The Whoppers of 2004," October 31, 2004 (www.factcheck.org/elections-2004/the_whoppers_of_2004.html).

SLEAZY ADS FOR ALL ETERNITY

There's probably a special place in hell for the people who dream up these kinds of political hatchet jobs, and maybe the punishment is having to watch your own sleazy ads and hear your own slimy speeches over and over again until "the last syllable of recorded time," as Shakespeare so elegantly put it. But whatever judgment these politicos face down the road, right now they're playing havoc with our ability to talk about the nation's real financial problems. And unless we stop listening, they'll keep us from fixing the problem while we still have time.

The Politically Dead and Departed

Back in the 1990s, H. Ross Perot launched a brief but notable political career urging Americans to take the federal budget deficit seriously.[14] Perot was a wealthy man who didn't have to raise money to run for office, so he had more freedom than most candidates to campaign on the issue that concerned him most. For a while, he attracted a sizable following, but his campaign eventually fell apart. Most observers think his downfall had more to do with Perot's personality than his position on the budget. "Prickly" and "eccentric" are words that often come up in connection with Mr. Perot,[15] but we've nominated him as one of our "Heroes of the Revolution" (see page 226).

---

[14] Steven A. Holmes with Doron P. Levin, "The 1992 Campaign: Independent Perot's Quest—a Special Report; A Man Who Says He Wants to Be Savior, If He's Asked," *New York Times*, April 13, 1992.

[15] See for example, "Hat in Ring, Head to Follow," *Rocky Mountain News* (Denver), October 2, 1995; see also "James Urges Republicans to 'Fight All the Pee-rows,'" *Birmingham News*, January 14, 1996.

Despite Perot's ability to attract a fairly significant political following by talking tough about the budget deficit, his experience is the exception. The conventional political wisdom—and it's pretty entrenched—is that candidates who want to win had better talk about cutting taxes and spending more money on things voters like—preferably both. Calling for higher taxes and/or cuts in popular programs like Social Security and Medicare is definitely not recommended. Recent political history is littered with some very promising candidates—established leaders with broad experience and strong backing—who fell by the wayside when they disobeyed "the rules."

Running for office is a complex, dicey affair, and there is generally more than one reason why a candidate wins or loses—money, personality, stands on issues, ability to get the vote out, cried during the campaign, didn't cry during the campaign, got a picture taken windsurfing, and many others. Yet anyone mulling a run for office has to consider the scary list of candidates who said the "wrong" things about taxes or spending and ended up making a concession speech when all the votes were counted.

This is where we the voters come in for our share of the blame. Why would we expect candidates to tell us the unvarnished truth given what generally happens when they do? It's really no wonder that most politicians today decide that their best bet is to avoid going out on the "how to balance the budget" limb.

Here's some of the political history that's gotten us where we are today.

★ **Vice President Walter Mondale** took what he himself called "a helluva shellacking"[16] in the 1984 presidential campaign

---

[16] Dan Balz and Milton Coleman, "Mondale Is Back, at Practicing Law; Defeated Candidate Admits Failings, Sees Signs of Vindication," *Washington Post,* April 8, 1985.

which he lost to President Ronald Reagan. Political columnist Tom Wicker summed up the campaign efficiently: "When Walter Mondale pledged a tax increase to cut the deficit, he lost forty-nine states—to a Republican President who had run up that deficit and doubled the national debt"[17] Mondale himself recognized what had gone wrong. Interviewed later, he said: "If you look at the campaign in retrospect, I looked like a person who was always talking about problems, about tough steps that were needed

Vice President Mondale said he would raise taxes to cut the deficit. He lost his presidential bid. *Credit: U.S. Senate Historical Office*

to solve problems. While my opponent was handing out rose petals, I was handing out coal."[18] Mondale's political demise was not lost on future presidential candidates. In fact, the headline on Wicker's column chronicling Mondale's defeat just about summed up the moral of the story politically speaking. "Death Wish and Taxes" is all most politicians really need to know.[19]

★ In this book, we give **President George H. W. Bush** (the first Bush) one of our "Heroes of the Revolution" awards for his role in the 1990 bipartisan budget compromise—one that set the nation on a more solid financial path for some years. But whatever the mer-

---

[17] Tom Wicker, "In the Nation: Death Wish and Taxes," *New York Times*, December 14, 1986.

[18] Balz and Coleman, "Mondale Is Back."

[19] Wicker, "In the Nation: Death Wish and Taxes."

The first President Bush suffered politically when he agreed to a budget compromise that included tax hikes. *Credit: Library of Congress, David Valdez, Photographer*

its of Bush's budget-minded decision making, he, too, got caught in the "death wish and taxes" quicksand. When he ran for office in 1988, he promised, "Read my lips, no new taxes."[20] When he got in office and faced up to some painful budget realities, he decided that tax hikes had to be part of the overall picture.[21] When the next election came round, President Bush became a one-term president despite having a 89 percent approval rating right after the Persian Gulf War.[22] President Bush also got caught glancing at his watch during a TV debate with Bill Clinton,[23] and his staff ended up assuring people that he had indeed been inside a grocery store "a year or so ago . . . in Kennebunkport," even though the president marveled at seeing the kind of price scanner commonly used at checkout counters at a manufacturing convention.[24] The

---

[20] Quotes of Note; the Best of '88, *Boston Globe,* December 31, 1988.

[21] Andrew Rosenthal, "Bush Now Concedes a Need for 'Tax Revenue Increases' to Reduce Deficit in Budget," *New York Times,* June 27, 1990.

[22] The Pew Research Center, "Modest Bush Approval Rating Boost at War's End; Economy Now Top National Issue," April 18, 2003.

[23] B. Drummond Ayres Jr., "The 1992 Campaign: Campaign Trail," *New York Times,* October 30, 1992.

[24] Andrew Rosenthal, "Bush Encounters the Supermarket, Amazed," *New York Times,* February 5, 1992.

man was just not having his best campaign. Still, there is little doubt that his decision to raise taxes while in office was a serious political wound.[25]

★ Some candidates get caught on the other side of the ledger. **Senator Bob Dole** tripped up on the spending side in his unsuccessful run against Bill Clinton in 1996. After the Clinton team fired a series of volleys suggesting that Dole would cut Medicare and raise Social Security taxes to pay for broad income tax cuts, the candidate was politically on the run. News reports described him as "brimming with frustration" in a West Palm Beach elderly center. "The last lady I talked to as I left," Dole complained, "this lady said, 'Why are you cutting my Medicare?' The lady pushing the wheelchair said, 'That's all she hears all day long, the Clinton ads, that you're going to cut Medicare, cut Medicare, cut Medicare.'"[26] Despite Dole's

Many voters feared Senator Bob Dole might cut Medicare. *Credit: U.S. Congress, House Committee on Veterans' Affairs*

claims that he, too, would protect programs for the elderly, Clinton trounced him in the election. Dole even ended up aggravating some prominent conservatives like David Frum, then at the

---

[25] Roberto Suro, "Viewing Chaos in the Capital, Americans Express Outrage," *New York Times,* October 19, 1990.

[26] Katharine Seelye, "Politics: The Republican; In Blistering Attack, Dole Says Clinton Is Using Scare Tactics," *New York Times,* September 27, 1996.

Manhattan Institute.[27] For Frum, "Dole's passionate declaration of his support for Medicare and Social Security only fortified the false notion that these programs are in fine shape." Frum complained that Dole, rather than advancing the debate on entitlement spending, had actually reinforced the political belief that Social Security and Medicare are untouchable. "There are worse things in politics than losing," Frum wrote.[28]

---

[27] Frum is now a resident fellow at American Enterpise Institue (www .davidfrum.com/aboutfrum.htm).

[28] David Frum, "The Big Scam of 1996," *New York Times,* November 6, 1996.

CHAPTER 13

Has K Street Become Washington's Main Street?

When we got into office, the thing that surprised me most was to find that things were just as bad as we'd been saying they were.

—President John F. Kennedy (1917–1963)

Election politics is probably the chief reason so many candidates and elected officials don't talk about the country's financial problems and haven't started to deal with them responsibly. It's a risky business to suggest raising taxes or spending less on something people like in order to solve a problem that seems far, far away. Even the feeblest call to hike taxes or cut benefits in Social Security or Medicare is likely to be a godsend to an election opponent.

But it's also obvious that politicians' reluctance to address this issue doesn't go away once the election is over. There is something about the way that Washington works these days that seems to fight against Congress and the administration tackling tough issues and making needed decisions about them. There is a boomlet of analyses from journalists, political scientists, and others examining why "Washington seems mired in dysfunction," as the *New York Times*'s John Broder phrased it. In fact, anyone who spends a lot of time watching the contemporary political scene could probably list dozens of malfunctions in the way the system works today. But in our little Budget Politics 101 introductory course, we're going to zoom in on just two of them—the ones we think could really hold us up on this particular issue.

The first is that members of Congress and the administration (from both parties) listen to lobbyists. The second is that good-old-fashioned bipartisan compromise does seem pretty old-fashioned these days. In our humble opinions, solving the country's budget challenges requires change in both areas. Decision makers need to remember that an assortment of special interests (represented by lobbyists) do not necessarily add up to the public interest. And they need to work with the other party and be willing to compromise. Let's take a look at what we face in each of these areas in the here and now.

THE K STREET CONNECTION

In Washington, lobbyists are affectionately known as "K Street" because so many of the big lobbying groups and firms have offices there. And there's no question that this lobbying business (and it's definitely a business) is really annoying to most Americans. Nearly eight in ten Americans say that political lobbyists have too much power and influence in Washington.[1] Both Republican and Democratic leaders have repeatedly promised to "do something" about it, but reforms seem to disappear or lose their teeth while the practice just keeps on growing.

In a devastating portrait of contemporary K Street,

---

[1] Harris Poll. February 6–12, 2007, from "The Polling Report" (www .pollingreport.com/politics.htm).

Jeffrey Birnbaum of the *Washington Post* reports that Capitol Hill now boasts roughly thirty thousand registered lobbyists, about double the number there were in 2000.[2] Showcasing the "industry's simple-yet-dazzling economics," Birnbaum describes one prominent lobbying firm, the Carmen Group, which brags "that for every $1 million its clients spend on its services, it delivers, on average, $100 million in government benefits."[3]

The pharmaceutical industry is the top spender, according to the Center for Responsive Politics, investing more than $900 million in lobbying between 1998 and 2005.[4] According to the *Washington Post*, industry lobbyists worked closely with Congress on the 2003 prescription drug plan, which provides taxpayer-supported insurance coverage for all seniors on Medicare and prohibits the agency from bargaining with drug companies for lower prices (we'll leave it to you to decide whether this is coincidental or not).[5] As Iowa senator Chuck Grassley, one of the bill's sponsors, described it, "You can hardly swing a cat by the tail in Washington without hitting a pharmaceutical lobbyist."[6]

Perhaps it's understandable that a major industry would want a presence on Capitol Hill while major legislation affecting it was up for debate. But even after the drug legislation passed in 2003, lobbying on health care issues has stayed at a fever pitch. In 2005, according to PoliticalMoney-Line (now CQMoneyline), a Web site focusing on lobbying,

---

[2] Jeffrey H. Birnbaum, "Washington's Once and Future Lobby," *Washington Post*, September 10, 2006.

[3] Ibid.

[4] R. Jeffrey Smith and Jeffrey H. Birnbaum, "Drug Bill Demonstrates Lobby's Pull," *Washington Post*, January 12, 2007.

[5] Ibid.

[6] Senator Grassley was quoted in ibid.

groups lobbying on health care spent more than $170 million in just the first half of 2005.

There's one other thing that makes this lobbying issue so complicated—these days, nearly everyone does it. Sure, a lot of lobbying benefits corporate, banking, and oil interests, but there are also lobbyists representing consumer groups, environmental groups, women's groups, organized labor, schoolteachers, higher education, and more.

American Indians, the Sierra Club, Goldman Sachs

Maybe you've forgotten the details of the Jack Abramoff lobbying scandal or never knew them in the first place. But however much you forgot or never knew, you might recall that Abramoff was accused of ripping off American Indian tribes who thought they were hiring a topflight D.C. lobbyist to help them.[7] Like many other groups in American society, American Indians often believe they need a Washington lobbyist to watch out for their interests.

To Principal Chief Jim Gray of Oklahoma's Osage Nation, "this isn't anything new; lobbying is an exercise in free speech. It just happens to be new for Americans to see Indian Country speaking its voice in this way."[8] And just in case you've leaped to the conclusion that most lobbying for American Indian tribes focuses on casinos and gambling, the main issues are actually "sovereignty, health, and housing," according to federal records examined by Oklahoma's *Tulsa Union* newspaper.[9]

---

[7] James V. Grimaldi, "Abramoff Indictment May Aid D.C. Inquiry; Lobbyist's Work for Tribes at Issue," *Washington Post*, August 13, 2005.

[8] S. E. Ruckman, "Tribes Put More into Lobbying," *Tulsa World*, August 28, 2006.

[9] Ibid.

Boldface Names

Lobbyists represent nearly all of us in some form or another,
and they include famous, accomplished people from across
the political spectrum.

- Former Texas governor Ann Richards, who died in 2006,
  was a lobbyist representing clients like Goldman Sachs,
  General Electric, and Martha Stewart Living Omni-
  media[10]
- Kansas senator and former presidential candidate Bob
  Dole became a lobbyist after losing his presidential bid,
  representing corporations like Tyco International and
  Johnson & Johnson.[11]
- Prominent Democratic spokesperson Anne Wexler, once
  an aide to President Jimmy Carter, is a partner of Wexler
  & Walker Public Policy Associates (the Walker is former
  Pennsylvania congressman Bob Walker). Her firm has
  represented American Airlines and General Motors.[12]
- Edgar Wayburn, five-term president of the Sierra Club,
  added D.C.-based lobbying to that environmental group's
  activities, thus convincing lawmakers to protect hundreds of millions of acres of American wilderness.[13]
- John Ashcroft, the former attorney general and Missouri
  senator, created the Ashcroft Group after he left the Bush

---

[10] Robert Elder, "Richards Turned to Lobbying, Reversing Role after
Governing," *Austin American-Statesman*, September 16, 2006.

[11] Matt Stearns, "Former Capitol Hill Legislators Cash In with Lucra-
tive Lobbying," *Kansas City Star*, December 21, 2004.

[12] "The Sharpest Shooters on K Street," *The Hill*, Capitol Hill Publish-
ing Corp., May 3, 2006.

[13] Julie Cart, "Conservationist Answered America's Call of the Wild,"
Los Angeles Times, September 17, 2006.

cabinet. An Israeli aircraft builder and eBay are among his clients.[14]

So what's a member of Congress to do when an election doesn't turn out his or her way? Even though House members aren't permitted to lobby their former colleagues for a year, or senators for two years, a fair number find their place in the wonderful world of K Street pretty quickly. They are allowed to advise and strategize during the moratorium period; they just can't contact legislators directly. Just months after the 2006 elections, defeated representatives Jim Davis (D-Fla.), Nancy Johnson (R-Conn.), Richard Pombo (R-Calif.), and Curt Weldon (R-Pa.), and former senator Conrad Burns of Montana had all found jobs in lobbying firms.[15]

Can We Rein It In?

Lobbying is a vexing problem for people who worry about government ethics and honesty, and, unfortunately, we can't cover the question adequately here. But it's worth mentioning the major new ethics law passed by Congress in 2007, designed to close off some of the connections between lobbyists and Congress—or at least make some of these back alleys a little better lit. Among the law's provisions:

★ Lobbyists are required to disclose when they organize "bundled" campaign contributions by getting multiple people to

---

[14] "Sharpest Shooters on K Street." See also "Ashcroft's Roster," *Washington Post,* August 12, 2006.

[15] Matt Kelley, "Ex-Lawmakers Find Work with Lobbyists; Federal Law Doesn't Forbid Firm Advisory Positions," *USA TODAY*, February 22, 2007.

contribute to the same candidate or party, at least when the contributions exceed $15,000 over a six-month period.

★ Members of Congress and their staffs are prohibited from accepting gifts (including meals) from lobbyists, and lobbyists are barred from knowingly giving gifts that violate congressional rules, with stiff penalties for lobbyists who break the law.

★ Candidates for the Senate and the presidency must pay charter rate for travel on private planes (previously, corporations could offer politicians a lift on the corporate jet and the pol had to reimburse only the price of a first-class airline ticket). House candidates must fly on commercial planes.

★ Members of Congress who leave and become lobbyists would have to disclose any job negotiations and potential conflicts of interest. Once a former member becomes a lobbyist, he or she loses the traditional ex-member's privilege of actually going onto the House or Senate floor to chat people up.

★ Members of Congress have to disclose all earmarks at least forty-eight hours before a vote (although legislative leaders are allowed to waive that rule). Members also certify that they and their families don't have a financial interest in the earmark.

★ New online databases will be set up for financial disclosures by lobbyists and members of Congress.

Is that going to be enough? Not likely. Oh, there's been what the *New York Times* called "a ripple of fear through K Street," as lobbyists bemoan the new penalties for picking up a check in much the same tone Dr. Smith used when interstellar travel went badly on *Lost in Space* ("Oh, the pain, the pain!").[16]

---

[16] David D. Kirkpatrick, "Tougher Rules Change Game for Lobbyists," *New York Times*, August 7, 2007.

And while making earmarks more public may head off some of the sneakier tactics, at this point it doesn't look like that's going to discourage lawmakers from using them. In fact, publicizing earmarks may have made the earmark market more competitive. Lots of members of Congress want full credit for bringing the bacon home to their districts (remember Senator Byrd). And they don't want people to see that some other member has been getting more for his district than they have for theirs. "Of course, when it becomes open to other members, everybody looks around and says 'Oh, I could have gotten that for myself,'" said Congressman Jose E. Serrano (D-N.Y.).[17]

So there's more to be done. To keep lobbying in line, some reformers have suggested changes like these.

- Go further in curbing campaign contributions from lobbyists and lobbying firms. For example, the watchdog group Public Citizen wants to limit contributions to less than $500 per election cycle and ban lobbyists from giving fancy parties to "honor" members of Congress, often at the political conventions.[18]
- Ban lobbying by members of lawmakers' families. Former Indiana senator Birch Bayh is a lobbyist; his son Evan is a sitting senator. For Indiana senator Richard Lugar, the situation is reversed. One of his sons is a lobbyist whose clients have included the Chamber of Commerce and Bank of America.[19] Former Democratic congressman Dan Mica

---

[17] Edmund L. Andrews and Robert Pear, "With New Rules, Congress Boasts of Pet Projects," *New York Times*, August 5, 2007.

[18] "Senate Has Failed to Deliver Effective Lobbying and Ethics Reform," *Public Citizen*, March 29, 2006, available at www.citizen .org/pressroom.

[19] Maureen Groppe, "Lobbying for Change in Congress; Proposals Made, but Lawmakers Can't Agree on How to Prevent Abuses, *Indianapolis Star*, January 22, 2006.

is a lobbyist, while his brother John is a sitting member of Congress, albeit on the Republican side.[20] The new ethics law imposes one important new restriction, requiring legislators to certify that earmarks do not benefit themselves or their immediate family. But earmarks aren't the only way decisions get made, and a lobbyist who's got a relative on the Hill is more likely to get a sympathetic hearing from other members.

- Make House lawmakers who leave government wait two years instead of one before becoming lobbyists, just like senators. Public Citizen calls this "slowing the revolving door."[21]

- Prohibit members of Congress from voting on issues in which they have investments or hold stocks, suggests law professor and commentator Jonathan Turley in an excellent op-ed for *USA TODAY*.[22] He suggests that members of Congress put their assets in blind trusts the way judges and officials in the executive branch have to do.[23]

- Ban earmarks entirely. Some significant steps have been taken. The 2007 ethics bill makes the earmark process more open. Also, when the Democrats took control of Congress in 2007, they imposed a one-year earmark moratorium, which Citizens Against Government Waste said cut congressional pork down from $29 billion in 2006 to

[20] Elizabeth Williamson, "Brothers Bridge Political Aisle," *Washington Post*, March 6, 2007.

[21] Public Citizen press release, "Lobbying and Ethics Reform Measure Produced Today Lacks Key Provisions," May 17, 2007, available at www.citizen.org/pressroom/release.cfm?ID=2439.

[22] Jonathan Turley, "A Question of Ethics," *USA TODAY*, November 14, 2006.

[23] Ibid.

$13.2 billion in 2007.[24] But that's not the same as stopping the practice. For example, some members of Congress have turned "phonemarking," or calling a federal agency directly to make sure earmarks from previous years stay in the current budget.[25] Basically, this means members can still swing money to a particular state or district. We will be surprised if Congress acts on the boldest of Jonathan Turley recommendations (and the one that most voters would probably support if they could)—ban earmarks entirely.[26]

- Create an independent ethics review committee that would include members of both political parties, plus some representative outside Congress. Fred Wertheimer, who has a long history of working with groups such as Common Cause and now Democracy 21, believes "there is a real need to get an effective and publicly credible system for enforcing the ethics rules. Right now, you have a non-credible ethics enforcement process that has failed overwhelmingly to do its job."[27]

Clearly, some of these ideas are worth thinking about, as are others, but not everyone is convinced changes like these will be effective. *New York Times* columnist David Brooks favors more disclosure about lobbyists' campaign contributions and a longer moratorium between leaving office and

---

[24] *2007 Congressional Pig Book Summary*, Citizens Against Government Waste, March 7, 2007, available at www.cagw.org/site/PageServer?pagename=reports_pigbook2007.

[25] "In the Democratic Congress, Pork Still Gets Served," *Washington Post*, May 24, 2007.

[26] Turley, "Question of Ethics."

[27] Elizabeth Williamson, "Democrats Lose Traction on Reforms," *Washington Post*, June 11, 2007.

assuming the lobbyist's mantel. But he thinks the results of try-ing to control the nice dinners out and flights on corporate jets will be negligible: "The bans on lobbyist-financed gifts, meals and travel are unimportant," Brooks wrote. "Few legislators are corrupted by a steak or even a ride in a Gulfstream."[28]

Frankly, we think government might improve some if elected officials had to foot the bill for their own steaks and sit in the middle seat back in coach with the rest of us. But even if lobbying could be scrubbed clean, having thirty thousand people running around Washington working hard for the "special" interest of the people who pay them means that the common interest of all of us can get lost in the pro-cess. Let's face it: how many of those thirty thousand people are out there lobbying for fiscal sanity?

We can all think of places where we want the govern-ment to spend more. We can all think of places where we'd like the tax laws changed so we don't have to pay as much. The problem is that when professional lobbyists begin pushing to spend more on this and ease taxes on that, the whole thing goes haywire. In some ways, it's like overfishing in the North Atlantic. From the point of view of every fish-ing boat captain, the best thing to do is to catch as many fish as possible, even though it's common knowledge that the fish stocks are being depleted. Unless the dynamic changes, unless some broader interest takes hold, people just keep doing what benefits them the most.

ENOUGH BIPARTISANSHIP TO FILL A THIMBLE

So is there any way to get elected officials to pay more atten-tion to the general interest and take on the important issues

---

[28] David Brooks, "Mr. Chips Goes to Congress," *New York Times*, Janu-ary 21, 2007.

with the tough choices? One way, at least historically, has been the grand bipartisan compromise. Both parties join in; neither gets exactly what they want; since elected officials on both sides of the aisle have to put their name to the less-than-perfect solution, they all get a little political cover. But based on national politics of the last decade or so, bipartisanship and compromise are beginning to look a little like the bald eagle—a noble American symbol that's on the endangered species list.

Once upon a time, Washington was considered "cozy" and "clubby." Now it's almost a cliché to call the current political climate "hyperpartisan" and "poisonous." And it would be easy to get into a long, angry, depressing argument over why this happened and whose fault it was. You could, if you wanted, follow it back to the disputed 2000 election or the Clinton impeachment or the confirmation hearings for a dozen all-but-forgotten political appointees, and, for all we know, the time that Stephen slapped Irene on *The Real World: Seattle*.[29] Tracing it all would take another book, and we don't have time for that now.

The important point for our purposes is that members of Congress used to be able to fight it out on the floor and during the campaign, and then sit down and cut deals on legislation over drinks afterward. It's not that Republicans and Democrats never cooperate on anything. The No Child Left Behind education law, the 9/11 Commission, the Iraq Study Group, and the 2007 immigration bill were all examples of some Republicans and Democrats getting together to do something together. But even bipartisanship doesn't seem to keep the peace or guarantee progress. These bipartisan efforts were and still are bitterly controversial.

"You can't reach across the aisle today," said former

---

[29] Not that we ever watched that show.

Republican senator Warren Rudman. "I am told that some members of Congress feel you're almost a traitor if you approach the other side to work with major legislation. People want to block that. Well, that is not the way most major legislation, domestic or foreign, has been made in our history. It's always been bipartisan. The amount of bipartisanship left on the Hill, you could put in a thimble right now."[30]

Not surprisingly, Washington's inability to focus on the public interest and solve problems most Americans really care about has soured public attitudes. A recent survey by the Pew Research Center showed that "Americans feel increasingly estranged from their government. Barely a third (34 percent) agree with the statement, 'most elected officials care what people like me think,' nearly matching the twenty-year low of 33 percent recorded in 1994 and a 10-point drop since 2002."[31] But we can't put all the blame on Congress and the president. The American public itself is giving mixed signals on its willingness to accept compromise. When the Pew Center asked Americans which kind of leader they admire most, "political leaders who make compromises" or "political leaders who stick to their positions," the compromisers won out, but only 51 percent to 40 percent.[32] As we've pointed out throughout this book, our leaders have been letting us down, but we haven't actually been doing our part as citizens, either.

---

[30] Panel discussion, Concord Coalition Economic Patriots Dinner, November 1, 2005 (www.concordcoalition.org/images/event-transcript/2005-patriotdinner/panel-discussion-transcript.html).

[31] The Pew Research Center, "Trends in Political Values and Core Attitudes: 1987–2007," Summary of Findings, March 22, 2007 (people-press.org/reports/display.php3?ReportID=312).

[32] Pew Research Center for the People and the Press, January 2007 News Interest Index, Final Topline, January 10–15, 2007 (people-press.org/reports/questionnaires/302.pdf).

So how can we begin to turn the tide? First, we urge you to check into and support the work of some of the groups trying to clean up politics. There's a list to start with in the appendix. Even more important—and it's basically the point of this whole book—we need to start demanding that elected officials take the country's financial problems seriously. We need to be prepared to hold them accountable when they don't. And we need to be realistic ourselves about what it will take to sort this financial mess out.

When our organization, Public Agenda, conducted focus groups on federal finances in 2006, most people grasped the magnitude of the problem after just a little discussion.[33] Most immediately saw that compromise would have to be part of the solution. Most were also pretty skeptical about politicians, wondering whether they would ever act, given all the incentives to continue with business as usual. At some point, however, someone in nearly every group would suggest that leaders in Washington would never let it really get out of hand because that would be so "insane."

That certainly would be insane. But that brand of insanity has struck governments before. The historian Barbara Tuchman wove a compelling and disturbing book *(The March of Folly)* out of examples of governments that stuck to policies contrary to their own best interests. Remember, there are powerful forces pulling on political leaders to skim over the budget problem. The "business-as-usual" mentality is very hard to fight. Unfortunately, just assuming that elected officials will wake up and smell the coffee on their own could be the most insane assumption of all.

Here's our goal for what needs to happen: every time an

---

[33] Public Agenda, *Facing Up to the Nation's Finances: Understanding Public Attitudes about the Federal Budget,* December 2006, available at www.publicagenda.org.

elected official sits down with a lobbyist to discuss what his or her client wants, and every time he or she is tempted to get on a "my way or no way" hobbyhorse, there should be a little voice in the back of their head asking, "What does this mean for the country's financial future? And what will voters at home do if I don't start acting responsibly on the country's budget problems?"

THE HEROES OF THE REVOLUTION

For thirty-one of the last thirty-five years, the country's annual budget has been in the red, but that doesn't mean that some people haven't been fighting the good fight. We'd like to salute a few of the individuals who have done their part to try to nudge the country to action.

SENATOR WILLIAM PROXMIRE

During a thirty-two-year Senate career, this Wisconsin Democrat was the "scourge of foolish federal spending," as *USA TODAY* put it.[34] From 1975 through 1988, Proxmire highlighted what he considered outrageous examples of government waste by presenting annual "Golden Fleece Awards" (for fleecing the taxpayers). Years later, the winners still make entertaining and relevant reading: $6,000 to prepare an

Wisconsin senator William Proxmire started the Golden Fleece Awards, highlighting government waste.
Credit: U.S. Senate Historical Office

---

[34] "American Originals," *USA TODAY,* December 16, 2005.

instruction manual for federal purchases of Worcestershire sauce;[35] $27,000 for research on why prisoners might want to escape,[36] and $20,000 for a model of the Great Wall of China in Bedford, Illinois.[37] Some critics considered the Golden Fleeces more hype than substance and complained that Proxmire routinely backed subsidies for dairy farmers in his own state.[38] But Proxmire was effective in calling attention to how Washington can fritter away taxpayer dollars, and for that we salute him. He died in 2005 at the age of ninety, but you can read the history of the Golden Fleece Awards and see a list of the Golden Fleece Top Ten at www.taxpayer.net/awards/goldenfleece (Taxpayers for Common Sense).

H. ROSS PEROT

We haven't heard much lately from H. Ross Perot, but in the 1990s, this Texas billionaire was the country's most famous deficit hawk. Running for president twice (in 1992 and 1996), Perot became famous for his infomercials on the national debt, complete with pie charts and himself as talking head,[39] and in 1992, running under the banner of the Reform Party, Perot got 19 percent of the vote, a historically high result for a third-party candidate.[40] There was "plenty to make you uneasy about Mr.

---

[35] Dave Umhoefer, "Obituary William Proxmire 1915–2005," *Milwaukee Journal Sentinal*, December 16, 2005.

[36] "American Originals."

[37] Umhoefer, "Obituary William Proxmire."

[38] "American Originals."

[39] CNN, "The Big Story; The Perot Factor," October 24, 1992; Hollace Weiner, "Perot Hammers at Old Themes in First TV Commercial of Campaign," *Fort Worth Star-Telegram*, September 2, 1996.

[40] Charles Krauthammer, "Bush: Two Great Challenges Met," *Washington Post*, November 23, 1992.

Ross Perot won the Distinguished Entrepreneur Award from the University of Southern Mississippi College of Business. We give him one for sounding the alarm about deficit spending. *Credit: Photo by Steve Rouse/University of Southern Mississippi*

Perot's ideas," according to editors at the *Washington Post,* but they gave him credit for "not ducking the number one serious issue."[41] Perot was cantankerous, sharp-tongued, erratic, and easily spoofed by Dana Carvey, and most Americans came to believe that he was not really presidential material. But he was a strong, forceful advocate for more fiscal responsibility, and if the country had heeded his warnings on the deficit and the debt, we might not be in such trouble now. So here's to you, Mr. Perot.

SENATORS PAUL TSONGAS AND WARREN RUDMAN

Paul Tsongas, a Democratic senator from Massachusetts, lost his party's presidential nomination to Bill Clinton and died of cancer in 1997.[42] Warren Rudman, a conservative

---

[41] "The Perot Advantage," *Washington Post,* March 31, 1992.

[42] Jack Beatty, Boston, MA; Noah Adams, "Tsongas Remembered," *All Things Considered,* National Public Radio, January 23, 1997.

Senators Paul Tsongas and Warren Rudman founded the Concord Coalition. *Credit: U.S. Senate Historical Office*

Republican senator from New Hampshire, wrote a book about his twelve years in Congress (1981–1993) and titled it *Combat*.[43] Yet the two joined former secretary of commerce Peter G. Peterson to found the bipartisan Concord Coalition, an organization that today is still fighting to call attention to the country's budget problems. When Tsongas and Rudman unveiled the Concord Coalition at a 1992 news conference in front of the "national debt clock" in Manhattan, Tsongas said "the American people are ready for the truth." Rudman added that "the time has come for citizens of this country to have another voice."[44] We couldn't agree more. We toast them both.

---

[43] Warren Rudman, *Combat: Twelve Years in the U.S. Senate* (Random House, 1996).

[44] "Coalition to Push Cutting of Deficit; Tsongas, Rudman Launch Grass-roots Effort to Trim Growing National Debt," *Boston Globe*, September 15, 1992.

PRESIDENT GEORGE H. W. BUSH

"Read my lips, I lied" is how the *New York Post* described the first President Bush's 1990 decision to join with congressional Democrats on a budget deal that included both spending cuts and tax increases.[45] The reference, of course, was to the president's quotable quote in the 1988 campaign, "Read my lips, no new taxes." Bush took heat from members of his own party for going back on his campaign promise and got very little admiration from anyone else. Normally, we don't applaud this kind of promise breaking ourselves, but keeping his pledge would have made the country's budget problems even worse. The 1990 tax increases were one of several important factors that helped the government balance its budget in the late 1990s. As columnist Jonathan Rauch wrote in the *New Republic*, President Bush could have done "the right thing politically or the right thing fiscally, but not both."[46] So when the choice is short-term politics versus the long-term interests of the American people, and you choose the public interest, you get our thumbs-up. President Bush no. 41, this is for you.

U.S. COMPTROLLER GENERAL DAVID WALKER AND THE FISCAL WAKE-UP TOUR

The comptroller general is the nation's accountant, and David Walker certainly looks the part. What's more, the man's got charts and statistics that put the fear of high school math into most of us. But in an era of rampant political waf-

[45] Larry Martz with Eleanor Clift, Thomas M. DeFrank and Ann McDaniel, "Biting the Bullet," *Newsweek*, July 9, 1990.

[46] Jonathan Rauch, "In Defense of Bush Senior," *San Diego Union-Tribune*, June 4, 2000, reprinted from the *New Republic*.

Comptroller General David Walker is the nation's accountant. He says our current financial path is "unsustainable." *Credit: U.S. Government Accountability Office*

fling in which lots of people shade the truth and try to spin past the inevitable, David Walker gets our admiration and respect. He's joined with a team of experts—Bob Bixby of the Concord Coalition, Alice Rivlin and Isabel Sawhill from the Brookings Institution, Stuart Butler of the Heritage Foundation, and others—on a speaking tour telling Americans there's trouble ahead unless we "wake up." In fact, it's called the "Fiscal Wake-up Tour," and if it comes to your town, it's definitely worth a couple of hours of your time (go to www.concordcoalition.org for a schedule). Walker closes his presentation of facts and figures with photos of several breathtakingly beautiful grandchildren. It's a reminder, he says, of what this is really all about. We can get together and fix the budget mess, or we can leave a poorer, more-troubled, and less-secure nation to our kids and grandkids.

When to Be Afraid, Very Afraid

It is easy to get confused about what they are up to in Washington when the billion- and trillion-dollar guesstimates combine with the "I have an advanced degree in economics, and you don't" jargon. But baffling statistics and expert gobbledygook are only part of it. Politicians have some very clever ways to cloud the issue while they push us deeper in debt. During elections, candidates tend to scurry around the hard facts as much as possible, but you need to keep an eye on politicians between elections, too. When it comes to the budget and the debt, our elected officials have not been doing us any favors lately. Here are the most popular ploys to look out for.

They promise popular new programs without saying how to pay for them. Americans seem to fall for this all the time, which is one of the reasons the nation is now about $9 trillion in debt and counting. Polls show most Americans want the government to spend more on health care, education, medical research, national parks, law enforcement, national security, and more. But unless something else is cut, or taxes are hiked to pay for this new spending, it will make the budget situation worse. We're not suggesting that the United States doesn't need to do more in some or perhaps all of these areas, but as citizens, we really have to start thinking about how to pay for what we want. There is no free lunch. Our kids are going to end up paying for this. It's just time to face up to that.

They promise tax cuts without saying where the money will come from. Taxes don't make people happy, especially when they apply to you personally, so cutting them is a staple of the campaign trail. Politicians who propose tax cuts may talk vaguely about "smaller government" or "making government more efficient" or "cutting waste and bureaucracy." Sometimes they say tax cuts will make the economy grow. The problem is that Congress passes tax cuts, the president signs the bills into law with a big flourish, and the government just continues to grow. As for the economy, any boost that tax cuts might offer now have to be weighed against threats posed by the huge amount of money we owe. Nearly every budget expert we could find—including well-respected conservative ones—says that we need to put the days of tax cuts without spending cuts to match behind us.

They put the costs in a "supplemental" spending bill. Sometimes Washington seems to operate on the same principle as many dieters: "if no one sees you eat that pint of chunky chocolate fudge with walnuts, the calories don't count." To avoid getting blamed for record-breaking deficits—and so they won't really have to tell the public how much things cost—politicians sometimes put major expenses in "supplemental budgets." Congress votes on these separately from the main budget (which is already huge), so it's harder for people to focus on the bottom line. Sometimes they string out really big budget items—like much of the money for the war in Iraq—into a series of supplemental appropriations: $70 billion here. $90 billion there; It's just so much easier than adding billions of dollars more to budgets that are already billions of dollars in the red.

They hide pet projects in "emergency" spending bills. When something terrible happens, a catastrophe like September 11 or Hurricane Katrina, Congress often passes emergency spending bills to address the problem. There is nothing wrong with this. Some things can't be predicted. We're a wealthy and compassionate country, and we want to help when our fellow citizens face tragedy and loss. But the definition of "tragedy" is elastic in Washington, as is the definition of "emergency." Hurricane Katrina was both tragic and an emergency, beyond doubt. A 2006 emergency spending bill to provide money for

Hurricane Katrina victims and the war in Iraq also contained money for higher education, relocating railroad tracks, and a "seafood promotion strategy."[47] You can decide for yourself whether these are good ways to spend taxpayer money, but it's hard to call them actual emergencies.

Sometimes aggressive reporting and a subsequent public outcry force politicians to back away from "now you see it, now you don't" budgeting and even renounce some of the odder spending projects tucked away in bigger pieces of legislation. Still, the takeaway message here is that you just have to watch these folks all the time.

---

[47] "New Criticism Falls on 'Supplemental Bills,'" *New York Times*, April 25, 2006.

★★★★★★★★★★★★★★★★★★★★★★★★★★★★★★

CHAPTER 14

2010–the High Noon of Budget Politics

Some debts are fun when you are acquiring them, but none are fun when you set about retiring them.

—*Ogden Nash*

No matter what you think of President Bush's 2001 and 2003 tax cuts (there were about a dozen different tax cuts signed into law), we may all eventually be grateful for one of their key features: nearly all of them expire at the same time. That means we have an opportunity to take a close look at our financial picture—and specific choices to make. The red-letter date to mark on your current events calendar? December 31, 2010.

According to projections by the Congressional Budget Office, extending all of President Bush's tax cuts would cost the U.S. Treasury roughly $1.8 trillion in the following decade.[1] Since the oldest baby boomers will be collecting

---

[1] Congressional Budget Office, "An Analysis of the President's Proposals for Fiscal Year 2008," March 2007, available at www.cbo.gov/ftpdocs/78xx/doc7878/03-21-PresidentsBudget.pdf.

Social Security and on the verge of getting Medicare by this time, the country's financial choices (if we're honest about them) will be grim.[2]

If we don't put more money into the system (raise taxes) and/or stop the money from flowing out (cut government programs and cut back on Social Security and Medicare benefits), the red ink will just hemorrhage from 2011 onward. This financial reality is not really in dispute. The Congressional Budget Office says it. The U.S. comptroller general, David Walker, says it. President Bush's former economic adviser Douglas Holtz-Eakin says it. Former secretary of commerce Pete Peterson says it. Former senators Warren Rudman and Bob Kerrey say it. Former Federal Reserve chairman Alan Greenspan says it, and so does current chairman Ben Bernanke. If we were making a speech at the Academy Awards, the music would come on, and we could still keep adding names of knowledgeable people who are sounding the alarm. Nearly everyone who knows anything about this issue says we have big, big trouble on the way unless we act.

DECEMBER 31, 2010–WHEN BUSH COMES TO SHOVE

President Bush will be chopping wood in Texas by this time (or maybe he'll be the baseball commissioner), but 2010 will be the "High Noon" of budget politics. Going cold turkey and allowing all of the Bush tax cuts to expire (and keeping the dreadful alternative minimum tax in its current form) would immediately bring additional money into the Treasury, increasing federal revenue by over 9 percent in 2011 and 7.5 percent

---

[2] Remember, Medicare kicks in at age sixty-five, but if you're willing to accept reduced benefits you can get Social Security at age sixty-two.

President Bush signed a series of tax cuts into law. Most expire at the close of 2010. *Credit: White House Photo by David Bohrer*

in 2012, "thereby bringing the budget into surplus "as the CBO puts it.[3] But there is very little in today's political chatter leading us to believe that "going cold turkey" is really on the table. In fact, in their 2007 budget resolution, congressional Democrats have already signaled they want to keep some of the Bush cuts, such as the child tax credit and the "marriage penalty" reduction.

We do expect that there will be plenty of political maneuvering and fancy dancing as elected officials try to avoid hard choices on the budget. We can confidently predict you'll hear some of the same old, same old political rhetoric—liberals insisting on rolling back the Bush tax cuts because they "just benefit the rich"; conservatives insisting on keeping all of them because "taxes are bad for the economy." We bet there will be some wearisome knockdown drag-outs over whether letting the cuts expire is actually a tax "increase" or not—complete with "yes, it is, no, it's not" screaming all over cable television. And since the actual decisions will revolve around tax brackets, tax credits, deductions, filing status,

[3] Statement of Peter R. Orszag, director of the Congressional Budget Office, to the Committee on the Budget, U.S. House of Representatives, "The Budget and Economic Outlook: Fiscal Years 2008–2017," January 30, 2007 (www.cbo.gov/ftpdocs/78xx/doc7878/03-21-PresidentsBudget.pdf).

percentage restrictions, and the "uniform definition of a child," a lot of us will get bleary-eyed reading the details. Some Americans will take a "tax cuts, good, tax increases, bad" attitude, and never rouse themselves to think about anything beyond that.

However, please believe us when we tell you that willful ignorance is the worst possible choice you can make. The years 2010 and 2011 are when we'll have our first good shot at this problem. This is the pitcher putting the ball right over the plate. If you don't take the bat off your shoulder and swing at this one, there may never be a better chance.

So we'll lay out the highlights here. By the way, we're giving you a very, very simplified introductory version. Much more detailed information and analysis can be found in the CBO's "Current Budget Projections,"[4] and at the IRS Web site at www.irs.gov (we know you just can't get enough of the IRS).

Choice 1: What should we do about income taxes?

When President Bush first ran for office in 2000, he was elected at least in part on his promise to cut taxes—and for most people, that means income taxes. Over the next few years, he made good on his word, signing legislation that reduced rates on individual taxpayers at different income levels, including those who earn at the very top levels. Republicans (and others) argue that these tax cuts have helped the U.S. economy by putting more money into people's hands to invest and spend, and that people from all walks of life have benefited. But given the country's mounting debt, and the problems we face on Social Security and Medicare, others want Congress to let these cuts expire and basically return to the tax rates we had under the Clinton administration. Democrats have

---

[4] You can find them at www.cbo.gov/budget/budproj.shtml.

generally criticized the Bush tax cuts as mainly benefiting the wealthy. When Senator John Kerry was running for president in 2004, he proposed eliminating these tax cuts for people earning more than $200,000 a year and using the money to expand health care coverage, among other things.[5] In the run-up to the 2008 presidential election, a number of Democratic contenders have also suggested eliminating the tax cuts on higher income earners and using the savings in other ways.

In 2010, the nation's choices may center on whether it's better to (a) keep these lower tax rates, (b) raise the taxes and use the money for other programs and services we want, or (c) raise the taxes and try to balance the books. Not an easy choice, we admit, but it's a discussion the country really needs to have with some frankness.

Choice 2: What about the estate tax?

As of 2007, you can inherit up to $2 million tax free, and inheritances to spouses are generally not subject to a federal estate tax at all.[6] Consequently, the estate tax currently affects fewer than 1 percent of American families.[7] Even so, about half of Americans (48 percent) say that they would be more likely to vote for a candidate who wants to repeal the estate tax, so the issue does have political currency, so to speak.[8] Those who want to repeal the estate tax permanently often

---

[5] Edmund L. Andrews, "Styles Similar in Bush and Kerry Duel on Deficit Numbers," *New York Times*, August 15, 2004.

[6] Tom Herman and Rachel Emma Silverman, "Republicans Consider Keeping Tax Alive for the Very Rich," *Wall Street Journal*, January 19, 2005.

[7] Edmund L. Andrews, "G.O.P. Fails in Attempt to Repeal Estate Tax," *New York Times*, June 9, 2006.

[8] NBC News/Wall Street Jounal Poll, June 9–12, 2006, from the Polling Report (www.pollingreport.com/budget.htm).

refer to it as "the death tax." They argue that the person who originally earned the money paid taxes on it in his or her lifetime, so it's not right to tax it a second time. People on the other side of the debate typically say the country shouldn't be offering another windfall for the very wealthiest Americans at a time when we have routine deficits, some 45 million Americans without health insurance, and when the U.S. is racking up major expenses fighting the wars in Iraq and Afghanistan. Some opponents of repealing the estate tax have started talking about the "Paris Hilton Benefit Act."[9]

Whether Congress decides to let the estate tax come back to life, or whether there's a move to repeal it permanently, may depend on which party has power. Democrats tend to see the estate tax as a fair way to raise money from wealthy heirs. Republicans tend to see this as a government taxing someone's hard-earned lifetime wealth yet a second time. One question raised by estate tax critics is how to handle inheritances that are actually small family businesses or farms (which can easily be worth a couple of million dollars), although an analysis by FactCheck.org suggests this is a relatively small slice of the issue.[10] Still, if Congress were in a mood to compromise, it might try to work out some ways to exempt inheritances in this category or reduce the tax's impact on them. But realistically, it's up to us whether Congress is of a mind to compromise or not.

---

[9] E. J. Dionne credits Michael J. Graetz and Ian Shapiro, authors of "Death by a Thousand Cuts," for coming up with the phrase in his column, "The Paris Hilton Tax Cut," *Washington Post,* April 12, 2005.

[10] Congressional Budget Office, "Effects of the Federal Estate Tax on Farms and Small Businesses," July 2005 (www.cbo.gov/ftpdocs/65xx/doc6512/07-06-EstateTax.pdf). See also "Estate Tax Malarkey: Misleading Ads Exaggerate What the Tax Costs Farmers, Small Businesses and 'Your Family,'" FactCheck.org, June 6, 2005 (www.factcheck.org/article328.html).

Choice 3: What about taxes for married couples?

Before the tax changes ushered in by President Bush, many Americans complained that the tax system was unfair to married couples—that there was a "marriage penalty" in the law.[11] To address this concern, the Bush tax package instituted two changes that are pretty helpful to married couples. One change raised the standard deduction for married couples and the other allowed low-income married couples to earn more before having to pay the lowest tax rate of 15 percent (you can check into the details at the IRS Web site).

On the face of it, you might assume that any change that reduced a "marriage penalty" would just automatically benefit middle-class taxpayers—at least the married ones. However, there was a healthy debate when this legislation was proposed about whom it was really helping and whether it was removing a "penalty" on marriage or giving a "bonus" to some married couples.[12] In 2001, for example, the standard deduction for married couples was 167 percent of the standard deduction for a single taxpayer.[13] In 2008, the standard deduction for married couples will be 190 percent of a single person's deduction and for 2009 it will go up to 200 percent. [14]

Can two live more cheaply than one? Should married people be able to pay taxes at the same rate they would if they

---

[11] See for example, Eric Pianin, "House Votes for Reduction in 'Marriage Penalty' Tax; Plan Covers Millions More Than Clinton's," *Washington Post*, February 11, 2000.

[12] Ibid. See also Iris J. Lav and James Sly, "Conference 'Marriage Penalty Relief' Provisions Reflect Poor Targeting," Center on Budget and Policy Priorities, July 21, 2000 (www.cbpp.org/7-21-00tax.pdf).

[13] Change from the Economic Growth and Tax Relief Reconciliation Act of 2001 as reported by the Tax Policy Center, Tax Facts, EGTRRA Marriage Penalty Relief Provisions.

[14] Ibid.

were still single? Should government try to make marriage a more appealing lifestyle through its tax policies? Is this fair or not? Come 2010, you may be hearing this debate again.

Choice 4: What about the higher tax credits for children and deductions for child care?

The tax code is filled with deductions and credits for this and that, but President Bush's tax packages included two that are especially helpful to families with children—a higher tax credit for each child and higher deductions for child care. The child tax credit is aimed mainly at middle- and low-income families; the credit is reduced for couples with incomes over $110,000 or single parents with incomes over $75,000.[15]

Families in which parents work and pay someone else to watch children also benefit from the higher deductions for child care. Again, the largest deductions go to those in the lower income brackets, but even more affluent families can deduct 20 percent of their child care expenses.[16] It does seem fairly hard-hearted to think of eliminating these tax benefits for families. After all, children are expensive. Most politicians will probably avoid this category like the plague when it comes to looking at tax hikes. Even suggesting reducing the deductions for child care for top-income people would probably result in a headline like "Senator So-and-so Calls for Tax Hikes on Families," so we don't expect much political push to change these.

Choice 5: What about tax rates on capital gains and dividends?

Most of the big tax battles involve legislation that expires at the end of 2010, but the tax law on on capital gains and divi-

---

[15] Internal Revenue Service, Publication No. 972, Child Tax Credit.

[16] Internal Revenue Service, Publication No. 503, Child and Dependent Care Expenses.

dends is set to expire at the end of 2008.[17] You might want to think of this as the skirmish before the big battle. President Bush had originally wanted to eliminate taxes on capital gains (the profit people earn when they sell property or stocks and bonds) and dividends (payments profitable companies make to shareholders). He believes these taxes discourage investment, which he sees as the powerhouse of a strong and growing economy. Democrats typically support higher taxes on capital gains and dividends since in their view the fact that you even have a capital gain or dividend is a sign that you're reasonably well-off. Congress didn't go along with the president's idea to eliminate this category of taxes, although in 2003 they did simplify the taxes (if you don't know, don't ask) and cut them substantially.[18] In 2006, Congress extended the lower rates until the end of 2008. When the House of Representatives passed the extension, Speaker Dennis Hastert said the vote signaled "a day of celebration for the American people."[19] Florida representative Alcee Hastings, a Democrat, complained, "Now millionaires have the right to have all the money they can," which gives you an idea of how controversial this particular tax area is.[20] Expect partisan fireworks on this one.

PUTTING THE BUSH TAX CUTS IN PERSPECTIVE

If you have this book in your hand along about March or April in any given year, chances are that the prospect of any

---

[17] "Phase-in and Expiration Schedule of Key Bush Tax Cuts under Current Law," Joint Economic Committee—Democrats, CBO Confirms That the Bush Tax Cuts Are Skewed Toward the Rich, August 2004, p. 5 (www.senate.gov).

[18] Ibid.

[19] Sheryl Gay Stolberg, "House Votes to Extend Investor Tax Cuts for 2 Years," *New York Times*, May 11, 2006.

[20] Ibid.

tax hike whatsoever will seem extremely unpleasant. But it may also be worth considering a little historic perspective. Even if all of President Bush's tax cuts went away, that would only put taxes back where they were in the 1990s, under President Clinton. On the other side, we have to be realistic about what can be accomplished if we just focus on the Bush tax cuts. According to CBO projections, even if Congress were to let all of the Bush tax cuts expire, the country would still face a big long-term budget problem. We still couldn't meet the commitments we've made on Social Security and Medicare. As Federal Reserve chairman Ben Bernanke testified to Congress in 2007, allowing *all* the Bush tax cuts to expire does "not come anywhere close to balancing the budget over the long run."[21]

THE BIG WRINKLE: WHAT SHOULD WE DO ABOUT THE ALTERNATIVE MINIMUM TAX?

There's another big tax decision we'll probably be making about this time, although perhaps we'll get to this one sooner. It's what to do about the alternative minimum tax (aka the AMT), a Byzantine set of tax rules originally passed in 1969 to ensure that very wealthy people wouldn't be able to weasel out of their share of taxes by using a lot of deductions. When the law was passed nearly forty years ago, it targeted people with incomes over $200,000, which would be about a cool million in today's dollars.[22] And that's the problem with

---

[21] Testimony of Ben S. Bernanke, chairman of the Federal Reserve Board of Governors, before the House Committee on the Budget, February 28, 2007 (www.access.gpo.gov/congress/house/house04ch110.html).

[22] David Cay Johnston, "It Doesn't Pay to Be in the A.M.T. Zone," *New York Times,* February 12, 2006.

the AMT. Unlike a lot of other parts of the tax code, it wasn't indexed to account for inflation, so now you have a lot of not exactly poor, but hardly wealthy, people paying this tax. Basically, if you have an upper-middle-class income, several children or multiple mortgages and live in a high-tax state, you might need to worry about this. The AMT is very complicated, and if you don't know the gory details personally, count yourself lucky. If you've had to pay it in the last couple of years, you're probably seething. Nearly everyone in and out of politics wants to fix the AMT, but putting this section of the tax code on a more rational footing will send red ink splashing every which way, like a Jackson Pollock painting of the tax code. According to the Congressional Research Service, just "patching" the AMT (basically fixing it for one year rather than fixing the basic problem with the law itself would cost about $60 billion—and that's just for one year.[23]

Debates about the Bush tax cuts sometimes skip over consideration of the AMT, but you shouldn't. We need to think through the whole thing together. Are there taxes we should raise, or, put another way, taxes we should return to the levels they were under President Clinton? Are there credits and deductions we need to reconsider? Are there taxes that should be lower for reasons of fairness or because of their effect on the economy? If so, what other taxes should be higher or what cuts should we make so that the numbers begin to balance out? Recommendation number one from our point of view—be flexible and prepared to compromise, because it is extremely unlikely that Americans will all see

[23] Congressional Research Service (Gregg Esenwein and Jane G. Gravelle), "CRS Report for Congress: Modifying the Alternative Minimum Tax (AMT): Revenue Costs and Potential Revenue Offsets," March 6, 2007.

these questions exactly the same way. Just ask a single person and a married person about the "marriage penalty," and see how much seeing eye-to-eyeing you get.

For us, our great hope is not that these issues get resolved according to our own personal preferences, but rather that our leaders will conduct the big tax debate of 2010 with an eye to the future, not just what seems most popular and convenient at the time.

★★★★★★★★★★★★★★★★★★★★★★★★★

CHAPTER 15

Tackling the Long-Term Problem One Bite at a Time

In the movie *Annie Hall*, Woody Allen sums up his philosophy of life with two dreadful old Borscht Belt jokes.[1] Luckily for you, we're only going to use one bad joke to explain what the country needs to do to tackle its federal budget problems.

The Question:	How do you eat an elephant?
The Answer:	One bite at a time.

Solving the budget mess is actually quite similar to having to eat an elephant—there's really no way to do it in one sitting. And regardless of what kinds of decisions we make

---

[1] Don't want to rent the DVD? Here they are: (1) Two old ladies are in a hotel in the Catskills. "Boy, the food at this place is really terrible," the first says. "Yes, I know," says the second, "and such small portions." (2) "I wouldn't want to belong to any club that would have someone like me for a member."

about President Bush's tax cuts—even if we cancel every single one of them—we still have a lot of work to do to get the country's long-term financial problems solved.

For most political leaders, the gargantuan size of the problem seems to provoke more artful dodging than genuine thought. And that's partly our own fault. Voter gratitude for taking even baby steps on this issue is practically nonexistent. The risks of suggesting a comprehensive deficit-and-debt reduction plan are downright scary to anyone who's chosen a career in politics. Offer up a list of concrete ideas—things to cut, taxes to raise—and you'll have so many people mad at you that your next campaign will be over before it starts.

But sooner or later—by choice or because we no longer have a choice—the country has to get moving on this problem, and it's likely to be a complicated, messy, lengthy, conflict-ridden affair. We didn't get ourselves into this situation in a year or two, and we're not going to be able to get out of it in a year or two either. Still, **if we break the problem down into smaller, manageable pieces, the country can make progress. But we have to be frank about what the pieces are and work up the nerve to take the first bite.**

As we said before, the country will face countless decisions on how to address the budget issue, and we'll have to make these decisions over and over again for many years. So what should we be debating? Where should we put our political energies? How do you even begin to think about a problem as big as this? Well, in fact, there are some choices about where to start—some backed broadly by experts, some more controversial, and at least some, in our view, that would be a bit of a detour. Here are the top contenders.

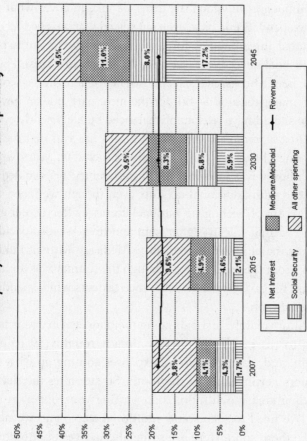

The Scary Projections on Revenues and Spending

2007: Net Interest 9.8%, Social Security 4.1%, Medicare/Medicaid 4.3%, All other spending 1.7%

2015: Net Interest 9.6%, Social Security 4.9%, Medicare/Medicaid 4.6%, All other spending 2.1%

2030: Net Interest 9.5%, Social Security 8.3%, Medicare/Medicaid 6.8%, All other spending 5.9%

2045: Net Interest 9.5%, Social Security 11.0%, Medicare/Medicaid 8.0%, All other spending 17.2%

Legend: Net Interest, Social Security, Medicare/Medicaid, All other spending, Revenue

The country will soon need to use nearly every tax dollar collected to pay for Social Security, Medicare, and interest on the debt. *Source: Government Accountability Office*

SHOULD WE GO COLD TURKEY ON DEFICIT SPENDING?

The one tasty little appetizer nearly all budget hawks recommend is going cold turkey on the federal government's addiction to deficit spending. They say the country needs to develop annual budgets in which the taxes we take in cover what we plan to spend. Some have even suggested that we need a constitutional amendment to force the federal government to balance its books every year (see page 26 for more on this). There might be an occasional exception—like when the economy is in a major recession—but for the most part, budget hawks say we should have reasonably balanced budgets nearly every year. Just a reminder, the United States has run deficits for thirty-one of the last thirty-five years, including years when the economy was growing. And as a country, we decided to go ahead with additional tax cuts even though we faced the added costs of beefing up national security after September 11 and waging war in Afghanistan and Iraq. We also decided to add a major new drug benefit to Medicare without talking much about the cost. So even though the country was already in debt, we gave ourselves some nice tax cuts while spending more and more.

Unfortunately, just balancing the budget every year isn't going to cut it. The country is already roughly $9 trillion in debt, and we have massive expenses coming up with the boomers getting long in the tooth. So stopping the deficit spending isn't much more than putting your finger in the dike. We need to take some additional steps, and we need to take them soon. Here are some alternatives about where to start first.

Fix Social Security First

Some experts say getting Social Security on a sound financial footing should be the first step. People generally understand how Social Security works. Millions of Americans depend on it (if for no other reason than to keep Grandma from moving in), so generating the political will to fix it should be doable. Guaranteeing the program's future might relieve a lot of the voter anxiety that makes the budget issue such a political hot potato. Compared to its sister program, Medicare, Social Security is in better financial shape, so many experts say the country should work on the "easier problem" first.

There are dozens of alternatives combining relatively acceptable cuts in benefits with relatively acceptable increases in taxes that could essentially solve the Social Security problem for a very, very long time. Plus, if the country could take this first step successfully, maybe we would be encouraged to move on to more daunting budget issues. The only hurdle here is whether the country can really negotiate a solution, or whether liberals who consider Social Security untouchable and conservatives who never liked the program in the first place will make compromise impossible.

Fix the Tax System First

Maybe you may think taxes are too high, or maybe you may think the country has gone overboard cutting taxes in recent years. But whatever you think about how *much* we pay in taxes, chances are you want tax cheaters to pay up and would love to see the tax system simplified. If the country really has to consider raising taxes to get its financial house in order, some people say we ought to fix the tax system first. We need to nab the cheaters and reduce the mind-boggling complexity of the tax code. We've actually heard of a few people who say they would be happy to pay even more in taxes if they just didn't have to spend the first half of April scurrying around the house looking for bank statements and rummaging through shoe boxes of receipts.

The IRS Board of Overseers, an independent group of experts that monitors the agency, suggested that upping the IRS budget by about 10 percent each year (about $435 million in 2006) would give it the muscle to collect about $1.74 billion in additional taxes.[2] That might seem like a good deal financially, but Congress rejected more modest increases in 2005.[3] More than a few Americans don't really like the idea of giving the IRS more money to audit, investigate, and prosecute tax cheating. More IRS audits might bring in more cash, but it's not hard to see why politicians might shy away from this in a campaign season ("Vote for Smith. He'll Boost Your Chances of Being Audited").

Making the tax code simpler is another "sounds good, but . . . " idea. Treating different kinds of income differently and having all sorts of deductions makes the system extremely complicated and cumbersome, but there are lots of people ready to scream their heads off if their favorite deduction is touched. In 2005, President Bush asked political powerhouses Senator John Breaux of Louisiana and former Florida senator Connie Mack to head up the President's Advisory Panel on Federal Tax Reform to devise a plan to simplify taxes. When the commission issued its ideas in late 2005, the U.S. Conference of Mayors objected to eliminating deductions for state and local taxes, [4] the National Associa-

---

[2] IRS Oversight Board, FY2006 IRS Budget Special Report, March 2005, available at www.treas.gov/irsob/documents/fy2006-budget-report.pdf.

[3] "Bush to Push for Bigger IRS Budget," Associated Press, February 1, 2005.

[4] U.S. Conference of Mayors Statements on the Report of the President's Advisory Panel on Federal Tax Reform, US Newswire, November 1, 2005.

tion of Realtors said proposals to change the home mortgage deduction would cause real estate prices to plunge,[5]
and the National Retail Federation said changes in the way
businesses deduct the cost of imported goods would in effect
mean a 30 percent tax increase for retailers.[6]

If you haven't heard much about the tax reform commission lately, that's because it was DOA (dead on arrival) in
Congress. Simplifying the tax system is chock-full of political land mines: touch the home mortgage deduction and
you die. And the mortgage deduction is only one of thousands of provisions in the tax code, all of them with vocal
and impassioned lobbies behind them. So you see how difficult this could get. Simplifying taxes might be a good idea in
concept, and maybe we should pursue it (and we talk more
about it ourselves in chapter 4), but the country could lose
an awful lot of time working on the budget problem while
all the various constituencies nuke it out over the tax code.
It's just a lot easier said than done.[7]

Get a Grip on Health Care Costs First

Right now, the U.S. government spends mega-dollars on
health care every year. Medicare pays for health insurance
and drug coverage for Americans over sixty-five. Medicaid
pays for health care for low-income Americans, and for
about half of the nation's nursing-home costs (for older
people who were often a lot better off financially before

---

[5] "Now the Debate Begins on Tax Reform," *Omaha World-Herald*,
November 5, 2005.

[6] Ibid.

[7] If you want to read what the advisory panel actually said, and perhaps observe a moment of reverent silence over its grave, the report is
available at www.taxreformpanel.gov.

they started needing twenty-four-hour medical care).[8] Plus the Department of Veterans Affairs provides health care for some 5.5 million vets, according to VA records, covering about 800,000 hospitalizations and some 60 million outpatient visits.[9] It's a lot of money now, and as time goes by, it's going to be a lot more.

There's a double-whammy problem here. The number of older Americans is increasing. Added to that, health care costs are rising faster than inflation.[10] According to a 2007 report from the Government Accountability Office, the problem of rising health care costs is so serious that it poses a threat "not just to the federal budget, but to American business and our society as a whole."[11] So some experts say doing something about the rising cost of health care should be our first plan of attack.

Again, this one definitely falls in the easier-said-than-done category. There are multiple reasons why health care costs are going up so fast and bitter fights among experts over what would help (enough for another book like this one just on health care costs, no doubt). Some people say the private system in the United States is inherently costly and want a system more like the Europeans have. Others counter that this would cause costs to skyrocket and quality to plummet. Some people say we need to ration very expen-

---

[8] Christopher J. Gearon, "Getting Stuck with the Tab; Tighter Asset Spend-Down Rules Will Force More Families to Cover Nursing Home Costs Alone," *Washington Post*, February 21, 2006.

[9] U.S. Department of Veterans Affairs, "VA Opening 38 New Clinics," May 29, 2007.

[10] Sabrina Eaton, "Baby Boomers to Face Higher Health Costs," *Cleveland Plain Dealer*, July 23, 2006.

[11] Government Accountability Office, "The Nation's Long-Term Fiscal Outlook," April 2007 update.

sive medical procedures, especially for the very old and terminally ill; others say this is both immoral and cruel. There are clashes over malpractice, the drug industry, how much financial responsibility people should shoulder themselves, whether technology and medical advances will drive costs up or hold them down by discovering less expensive ways to help people. And so on it goes. Given all this, the prospects aren't good for quick and easy wins on reducing what the government spends on health care.

Nonetheless, very few experts think we can really do much about the budget if we ignore this arena entirely. What's more, as the GAO report underscores, rising health care costs are endangering businesses and individuals as well as government. Just to get your attention, the AARP's John Rother suggests that all of us put aside a couple of hundred thousand dollars in our retirement plans just to cover our personal health care costs when we're old.[12] If that's the case, we'd all better acquire a taste for cat food.

Improve the Congressional Budget Process First

If you remember those "how a bill goes through Congress" charts from junior high, you might think the process Congress uses to decide how to spend taxpayer dollars is logical and straightforward. But not in today's Washington. Congressional rules and procedures regarding the budget are elaborate, convoluted, and slippery to boot.

It's hard for most of us to keep track of what actually happens to our money given all the clever ways Congress has to confuse things. That's bad enough, but the process has gotten so bad that even *Congress* doesn't actually know how much it's spending. For example, most of the cost for Iraq wasn't

---

[12] Eaton, "Baby Boomers to Face Higher Health Costs."

included in the regular budget for the first several years of the war (see "The Fortunes of War" in chapter 5). The requests for money for the war were submitted in "supplemental" spending bills that bypass the usual appropriations process. War costs are also spread out in various places in the defense budget, which, as the Iraq Study Group points out, means you have to be a professional budget analyst to figure out how much the war costs. And maybe you can't figure it out even then.

Then there are those tasty little earmarks—tasty and "little" to people in Washington, that is. In 2006, Congress tucked more than thirteen thousand of these custom-tailored spending items into larger, more important bills.[13] Generally, they benefit someone or some group in the representative's home district—like the people starting the teacup museum in North Carolina. In the past, earmarks weren't debated openly the way spending on national security or national parks would be. Members of Congress courteously allowed each other to bring a little pork home each year at taxpayer expense without making a fuss about it. Altogether, earmarks added nearly $50 billion to the 2005 budget, according to the nonpartisan Congressional Research Service.[14] There has been a lot of discussion about reining in earmarks, and in 2007, Congress passed legislation that makes it easier to know who's sponsoring earmarks and why. Even so, like bacteria that mutate and no longer respond to antibiotics, a lot of earmark-type spending seems to be slithering through anyway.[15]

The question is whether Congress can make good deci-

---

[13] John Solomon and Jeffrey H. Birnbaum, "In the Democratic Congress, Pork Still Gets Served; 'Phonemarking' Is Among Ways Around Appropriations Process," *Washington Post,* May 24, 2007.

[14] "Hooked on Handouts," *USA TODAY,* August 1, 2006, p. 12A.

[15] Solomon and Birnbaum, "In the Democratic Congress, Pork Still Gets Served."

sions given the Byzantine process it has now. There are plenty of ideas for fixing the problem—and some of them are almost as complicated as the processes they would replace. Check out David Bauman's explanation of a recent reform plan that included "line-item enhanced rescission, budget caps, sequestration, biennial budgeting, commissions, the kitchen sink, 2007 draft picks and a provision to be named later" to see what we mean.[16]

But there are some relatively straightforward steps many budget experts think are essential. Put everything in the budget; stop hiding things away in "supplemental" or "emergency" spending bills. Get rid of the earmarks—bring all spending out in the open and subject it to debate. Publish a complete list of every spending item Congress approves and every member who voted for it and put it on the Internet for easy access (it's called "transparency"). Adopt an enforceable "pay-as-you-go" system where Congress can't add new spending without raising taxes or cutting existing programs to pay for the addition. A real pay-go system would also prevent Congress from cutting taxes without cutting spending enough to stay out of the red ink.

Are changes like these blindingly obvious? Yes. Are they necessary? We think so, and for two reasons. One is that they probably would help ensure that taxpayer dollars are used more carefully. And second, if the rest of us are going to have to make some sacrifices, we think it's only fair that Congress clean up its act, too.

Clean Up Politics First

When we were doing the research for this book, we went down to Wilmington, Delaware, to attend a session of the Fiscal Wake-up Tour. The Wake-up Tour is a traveling road show

---

[16] "Budget Reform, Yet Again," *National Journal*, July 21, 2006.

that includes panel presentations from Comptroller General David Walker, budget experts from think tanks like the Heritage Foundation and the Brookings Institution, and representatives from the Concord Coalition. Each of the presenters talked about the crisis the country could face if we don't begin to face up to some financial realities. These people know their stuff. It's a powerful presentation, and you should catch it yourself if you can (visit www.concordcoalition.org to see where it's playing).

When the time came for audience Q & A, we were surprised by the number of questions that focused on how to fix the political system rather than on ideas for cutting spending or raising revenues. Should we reform campaign financing? What about curbing lobbying? What about federal funding of campaigns? What about term limits?

What people in the audience seemed to be saying is this: If we have to do things that are not so pleasant—pay higher taxes, postpone our retirements, cover more of our own health care costs, and so on—what can we do to ensure that the people who represent us in Washington really have our best interests at heart? How can we be sure that the solutions they vote for are fair?

This is a tough one. If you've been reading along with us, you know we're not happy with what's been going on in Washington. As a group, our elected officials really haven't been watching out for our interests. It's certainly hard to argue that the political system as it is now is acceptable. It's pretty awful, frankly.

But the question is whether the country can really afford to fix the political system *first*. Every year we lose tackling the budget issue makes the problem even worse. The debt gets higher and the time to fix Social Security and Medicare gets shorter. GAO reports are generally pretty dry, but here's what one has to say about putting off the financial choices: "The longer action to deal with the Nation's long-term fiscal outlook is delayed, the greater the risk that the eventual changes will be disruptive and destabilizing."[17] Scary words; words we shouldn't ignore.

What's more, people have been trying to fix up the political system for years, proposing ideas like term limits and federal funding for campaigns without much happening. Campaign finance reform is a very squiggly thing—you fix one problem, and new ones break out somewhere else. We're not saying that we think the country should just live with the political system we have now. And the country can certainly work on different problems at once. In fact, we have to—the country still has to address terrorism, inequality, energy independence, the environment, and other extremely important areas. We can't let those slide while we stop to debate the budget.

But everything we have seen and learned while working on this book leads us to one conclusion: We can't wait another year to get going on the budget mess. So saying that we have to fix the political system first could turn out to be a gigantic mistake.

---

[17] Government Accountability Office, "Nation's Long-Term Fiscal Outlook."

So Has Anyone Got Any Bright Ideas?

There are lots of fiscal experts who say the way the U.S. government approaches budget issues is all wrong. In fact, the U.S. government doesn't really have a long-term fiscal plan, and our year-by-year process is a little loose as well. It's hard to believe, but there's no single moment when Congress votes on "the budget" for the year.

By law, the president has to submit a budget request to Congress every February, which usually includes goals and projections (recently for five years out). And while the president's request sets up much of the budget debate, Congress doesn't have to accept a single thing in it. Instead, Congress passes its annual budget resolution, which is nonbinding (and thus not subject to a presidential veto) but sets the structure for what Congress will do with federal finances that year.

But neither the president's request nor the congressional resolution is the actual budget. There are no less than thirteen separate spending bills that Congress has to pass to fund various departments and agencies, from Agriculture to Veterans Affairs. Those bills are each debated separately and may be vetoed by the president. Entitlements don't count in this. Unless Congress makes a specific decision to pass legislation affecting Social Security and Medicare, the programs clip along on autopilot without Congress or the president doing anything.[18]

Obviously, this isn't how corporations handle their budgets (although Enron and WorldCom prove there are plenty of ways to get fiscally funky in the private sector). But even in the realm of government finances, is there a better option?

The European Union does have broad outlines for member countries. To be an EU member, a country has to agree to keep budget deficits under 3 percent of the country's gross domestic product and the public debt at 60 percent. Amazingly, by that standard, the United States does quite well, with current deficit at 1.6 percent of GDP and debt at about 37 percent. As we said, the EU's guidelines are very broad. And even with those guidelines, there's a procedure for making exceptions.

---

[18] Heniff, Bill Jr., "Overview of the Congressional Budget Process," Congressional Research Service, October 21, 1999.

SHOULD WE GO DUTCH?

A lot of fiscal experts like the way the Netherlands handles its budgets. Dutch governments are generally coalitions of competing parliamentary parties, and as part of putting together the coalition, the parties hammer out a budget blueprint that covers the entire four years until the next election. The blueprint includes caps for spending and rules for changing the budget if tax revenue comes in higher or lower than expected. If the budget runs a surplus, the parties agree on what percentage will go to tax cuts or paying off the national debt. Politicians being politicians, this system isn't perfect, but it does have a certain logical appeal. Given the bitterness of our own politics, the idea of Republicans and Democrats sitting down to agree on a four-year plan for what they will spend seems like a political pipe dream. On the other hand, both the first President Bush and President Clinton managed to hammer out bipartisan budget deals when they were in office, so maybe we shouldn't toss this idea out so quickly.

DO AS THEY DO IN NEW ZEALAND?

New Zealand has a different strategy, which is based on setting goals and being clear and open about the results of government policies. By law, the New Zealand government has to set long-term fiscal goals (like keeping the national debt at a reasonable level), and every year the government is required to report how it's doing in meeting these goals. Just prior to elections, the government has to report on how campaign promises will affect the budget (yes, we're trying to imagine that, too). Again, the system is not perfect, but New Zealand does consistently run budget surpluses.[19]

TIME FOR A BALANCED BUDGET AMENDMENT?

One option that's often been proposed is a constitutional amendment requiring a balanced budget. Forty-nine states have some

---

[19] Allen Schick, "Can the U.S. Government Live Within Its Means? Lessons from Abroad," Brookings Institution Policy Brief, June 2005.

sort of provision that their state budgets should balance. But this has never really gained a political foothold on the federal level. And it all depends on how the amendment is crafted. The state balanced-budget rules often look good on paper but aren't that rigid in practice. Many states have rules on the books but no provision for enforcing them.[20] And lots of state budgets over the years have been "balanced" using one-shot gimmicks and other short-term schemes that meet the law but don't address the underlying problems.

WHAT ABOUT THE LINE-ITEM VETO AND PAY-GO?

Another possible constitutional amendment would give the president a line-item veto. Again, this is a tool many state governors have that allows them to strike out portions of a bill while approving the rest. The president can only sign or veto entire bills, warts and all. A line-item veto would certainly give the president the power to strike out pork-barrel projects—or, at least, the ones that don't benefit the president's party.

Other changes don't require amending the Constitution. We've already talked about "pay-go" rules, which basically require that if Congress wants to spend more money on a program or cut a tax, it has to make up that lost money by cutting programs or raising revenue elsewhere in the budget. Pay-go legislation was a key tool in balancing the budget in the 1990s, but it was allowed to expire in 2002.[21] Currently, both the House and Senate have built pay-go into their legislative rules, but they can be pretty broad-minded in interpreting them. For example, the rules allow Congress to assume the Bush tax cuts expire as a way of offsetting new spend-

---

[20] Ronald K. Snell, "State Balanced Budget Requirements: Provisions and Practice," National Conference of State Legislatures, www.ncsl.org/programs/fiscal/balbuda.htm, accessed May 19, 2007.

[21] "Federal Budget Process Reform in the 110th Congress: A Brief Overview," Congressional Research Service, January 22, 2007.

ing. In other words, Congress can spend more money now if there's more revenue in 2010. Except, of course, that money may not arrive—and then what do you do? Congress can choose to waive pay-go rules in certain circumstances—and the Concord Coalition says the budget resolution for 2008 seems to assume a wavier to permit extension of some tax cuts.[22]

All of these budget strategies, at home and abroad, are no better than the politicians who implement them. The good news is that the raw material is there to turn this situation around. The Congressional Budget Office and the Social Security and Medicare trustees provide honest assessments of the nation's finances. And the bipartisan will to change has emerged before—not that long ago, in fact. The government balanced its budget in the 1990s in large part because President George H. W. Bush and a Democratic Congress were willing to raise taxes, and because President Clinton and a Republican Congress were willing to maintain fiscal discipline.

We can get the nation's finances under control. It's just going to require the determination—and a lot more scrutiny from the public—to get the job done.

---

[22] "Concord Coalition Applauds Paygo in Budget Resolution, but Warns That Projected Surplus Requires Hard Choices," press release, May 17, 2007, available at www.concordcoalition.org/new-publications.html.

CHAPTER 16

OK, if You're So Smart . . .

By now, you've got a pretty good idea of what the country's facing. You've probably got some ideas what should be done, too. So let's see what you can do about it. People used to say any American child could grow up to be president. Well, in this chapter, that means you.

The people on the Fiscal Wake-up Tour—that joint effort by both conservative and liberal budget experts to get people to pay attention to this problem—are fond of Comptroller General David Walker's best line, which is "when you find yourself in a hole, the first thing is to stop digging."

That's a great quote, and has a lot of truth in it, but it's also slightly misleading. If you're digging a hole in the ground, and you decide you don't want to go any deeper, you just put down the shovel. It's a passive choice, even a restful one, since digging a hole in the ground is hard work. It implies that to stop digging is easy and that the hard part is climbing out of the hole again.

The fact is, when you're talking about the federal budget, *the hole gets deeper when you sit on your hands and do nothing.* Every year the government runs a deficit, Washington has to borrow money, the debt gets bigger, and the hole gets deeper. And running a deficit is the default setting. It's easy. Balancing the budget takes effort. Even trimming the deficit down a little takes effort. Leaning on your shovel makes things worse.

Unfortunately, attacking this problem is a lot more like losing weight (we find that as depressing an analogy as you do). You've got to work at it, eat less and exercise. And just like deficit reduction, there are a lot of people out there pushing fad diets in which you can somehow eat anything you want, watch TV all day, make no sacrifices of any kind, and still lose twenty pounds. In both cases, it's not going to happen.

Think of the worksheet in this chapter as your chance to be both president and gym trainer. In fact, you get to be somewhat more powerful than either the president or your gym trainer, because (a) you don't need to negotiate any of your choices with Congress and (b) whatever you decide actually gets done—there's no going home from the gym and eating an entire pie.

The goal of the exercise is to stop digging the fiscal hole by trying to balance the federal budget in a given year. We've given you a list of programs and revenue sources. You can increase, decrease, or eliminate any item you want, but you are limited to those "discretionary" programs that can vary from year to year. (There are, as we've discussed, a lot of things that could be done with entitlements, but very few of them pay off right away, so they don't affect a one-year budget). A few things are off limits—you can't cut interest payments on the debt, for example. The banks have to get their money.

In the real world, of course, the long-term entitlement problems are the bigger concern, but controlling the year-to-year budget is a good start. Also in the real world, there are political and economic consequences to any of these decisions. We've pointed them out in the worksheet. You can factor them into your decisions, or not, as you please. But you should realize that the real president and Congress are going to be weighing these consequences, whether you do or not.

PROGRAM In millions of dollars Minus indicates revenue	2006 BUDGET	YOUR BUDGET	PROS	CONS	WHO CARES
Agriculture	**2006 Total:** $25,970				
Farm income stabilization (aka, crop subsidies)	$21,411		This has kept a lot of family farms in business.	Most of this money now goes to big agribusiness, which can take care of itself.	The farm lobby. Don't underestimate them.
Research and services (including inspections)	$4,559		Research helps keep us the world's breadbasket, and there has to be a watchdog over the nation's food supply.	Greater self-regulation would be cheaper, and the food industry knows it has to keep its act clean.	The food industry, along with anyone who eats
Community and Regional Development	**2006 Total:** $54,531				
Area and regional development, including rural/Indian programs	$2,580		Promotes economic growth in some of the poorest areas in the country.	There's a serious question whether these programs are effective as is.	People in poor, rural areas
Disaster relief and flood insurance	$46,106		Helps out people suffering from flood, fires, and general "acts of God."	Flood insurance, in particular, may encourage people to rebuild in dangerous areas instead of moving somewhere safer.	Disaster victims, anyone living on the coasts, and everyone who wants people to get help in a crisis
Community development	$5,845				

PROGRAM In millions of dollars Minus indicates revenue	2006 BUDGET	YOUR BUDGET	PROS	CONS	WHO CARES
Commerce and Housing Credit (a real grab-bag category)	**2006 Total: $6,188**			Overall, a lot of these functions could be turned over to the private sector.	
Mortgage credits	−$619		Helps low-income folks get homes		Homeowners, plus the building and banking industries
Postal service	−$971		Don't you want your mail?		Everyone who gets mail, plus the postal unions
Deposit insurance	−$1,110		Guarantees you don't go broke just because your bank does		Everyone with a bank account
Other "advancement of commerce" including refugees, low-income heating, etc.	8,888		All kinds of useful stuff: economic statistics, patents, small business loans, international trade help, etc.		The business world in general
Education, Employment, and Social Services	**2006 Total: $118,560**				
Elementary, secondary, and vocational	$39,710		Helps local schools offer better programs, particularly in special education and to meet No Child Left Behind goals	Schools have always been a local responsibility—federal aid is small compared to the state and local money involved	Parents, teachers

PROGRAM In millions of dollars Minus indicates revenue	2006 BUDGET	YOUR BUDGET	PROS	CONS	WHO CARES
Higher Education	$50,471		Helps millions of people go to college	We should encourage more people to save for college, not to expect government aid.	All the middle-class families expecting to send their kids to college
Research and general education aids (Library of Congress, Smithsonian, Public Broadcasting)	$3,076		Some of the best-loved institutions in the country	Public Broadcasting, in particular, draws heavily on private funds.	Kermit the Frog, who has many supporters, not to mention teachers, book lovers, and people who love shows with British accents
Training and employment	$7,199		A key element in helping displaced workers adjust and welfare recipients into jobs	There's mixed evidence on whether job training actually works. Plus, there are programs scattered around multiple agencies.	The unemployed
Social Services	$16,473		Federal support for local programs helping children, the elderly, and the disabled	This subsidizes local programs, which may be worthy but could also be funded locally.	Children, the aging, and the disabled, plus everyone who cares about them

PROGRAM In millions of dollars Minus indicates revenue	2006 BUDGET	YOUR BUDGET	PROS	CONS	WHO CARES
Other labor services (statistics, law enforcement, etc.)	$1,631		We need to know the unemployment rate, for example, and enforce health and safety laws.		Economists, workers, unions
General Government	**2006 Total:** $18,215		Somebody's got to pay for this stuff. Buried in there are some significant items, like the IRS, White House operations (including antidrug campaigns and OMB) and of course Congress itself.	There's always something to cut here—lots of people could do with fewer limos. Plus, a simpler tax code could mean we'd spend less on collections.	Enough said
Congress and legislative functions	$3,446				
Executive direction and management	$522				
Central fiscal operations	$10,165				
Center personnel management	$151				
Property and records management	$328				

PROGRAM In millions of dollars Minus indicates revenue	2006 BUDGET	YOUR BUDGET	PROS	CONS	WHO CARES
General purpose fiscal assistance	$3,798				
Other general government	$1,164				
Offsetting receipts	−$1,359				
Health	**2006 Total: $252,780**				Anyone who ever gets sick, eats, or buys an appliance
Health care services (public health, substance abuse, mental health, disease control)	$220,841		Fights just about every disease you care to name, and you'll be glad of them should pandemic flu ever strike		
Research and training (National Institutes of Health, clinical training, etc.)	$28,828		Conducts medical research on nearly everything from cancer to cankers		There's a plethora of groups who lobby for research on specific diseases (often at the expense of others).
Consumer and occupational health and safety (FDA, OSHA, Consumer Product Safety Commission)	$3,111		Keeps food, workplaces, and household products safe	Can be hidebound and slow to both approve good products and stop bad ones	Consumer groups and business

PROGRAM In millions of dollars Minus indicates revenue	2006 BUDGET	YOUR BUDGET	PROS	CONS	WHO CARES
International Affairs	**2006 Total:** $29,549				
Development and humanitarian aid	$16,720		Helps those in the poorest nations on Earth	Critics say this money is often wasted or ineffective.	Foreigners and foreign policy experts who say you might need that goodwill down the road
Security assistance (including military aid)	$7,811		Helps our allies defend themselves (and us, by extension)	Often goes to regimes with nasty human rights records	
Conduct of foreign affairs (mostly embassies and diplomatic operations)	$8,568		If you're going to negotiate with people, you need negotiators		
Foreign information and exchange	$1,176		Builds a positive image of the U.S. abroad—a useful thing to have	This hasn't been working so well lately.	
Financial programs	$4,726				

PROGRAM In millions of dollars Minus indicates revenue	2006 BUDGET	YOUR BUDGET	PROS	CONS	WHO CARES
Justice System	**2006 Total:** $41,016		Law and order is fundamental to any society.	You can argue whether lots of these efforts are winnable and cost-effective, (like the war on drugs) or whether some federal crimes should be handled by local police.	Everybody cares about crime, and a society where no one upholds the law is no society at all.
Federal law enforcement (including FBI, DEA, ICE, Homeland Security, ATF, IRS enforcement, Secret Service)	$20,039				
Federal prisons	$6,158				
Criminal justice assistance (victims benefits, pensions, aid to local agencies)	$4,768				
Federal courts/ litigation	$10,051				

PROGRAM In millions of dollars Minus indicates revenue	2006 BUDGET	YOUR BUDGET	PROS	CONS	WHO CARES
Natural Resources and the Environment	**2006 Total: $33,055**				
Water (Corps of Engineers, Bureau of Reclamation)	$8,026		Keeps rivers and harbors navigable	Critics say many flood-control measures actually make things worse, and nature should take its course.	Anyone on the coast, the shipping industry
Conservation and land management (Forest Service, Fish and Wildlife, public lands)	$7,813		Preserves vast areas of public lands, especially in the West	Get hit from both sides— business says there isn't enough exploitation of these lands, environ-mentalists say there's too much.	Governors, who'll have to find the money to do this if it's cut
Recreational resources (national parks, landmarks)	$3,069		Staggeringly beautiful, perhaps the greatest tourist attractions in the world	Tourists could be paying more of the freight with higher fees	Vacationers
Pollution control	$8,572		Critical to quality of life and protecting the environment	Too bureaucratic and rulebound for many in business	Anyone who cares about pollution— i.e., almost everyone
Other resources	$5,575				

PROGRAM In millions of dollars Minus indicates revenue	2006 BUDGET	YOUR BUDGET	PROS	CONS	WHO CARES
National Defense	**2006 Total:** $521,840		Most of this budget comes down to a debate about whether we have too much military or not enough. The U.S. outspends every other country on Earth on defense, yet still finds itself overstretched in Iraq and elsewhere. You can increase the budget or curtail the missions, but despite the considerable waste documented here, this is mostly about policy choices.		
Personnel	$127,543				
Operations/ maintenance	$203,789				
Atomic energy defense activities	$17,468				
Procurement	$89,757				
Research and development	$68,629				
Family housing	$3,717				
Other	−$-370				
Military construction	$6,245				
Defense-related activities	$5,062				
Net interest	**2006 Total:** $226,603	$226,603	This is the cost of doing business—the government needs to borrow money.	As we've mentioned, this is going to get out of hand if nothing is done.	Right now, just the banks—if things don't change, we'll all start caring about this.
Science, space, and technology	**2006 Total:** $23,616		Generally, advances human knowledge and provides untold long-term practical benefits	For many, this is in the "nice-to-have" category, particularly given earthbound problems.	Scientists. engineers, trekkies

PROGRAM In millions of dollars Minus indicates revenue	2006 BUDGET	YOUR BUDGET	PROS	CONS	WHO CARES
Space flight and research	$14,491				
General science and basic research	$9,125				
Transportation	**2006 Total:** $70,244				
Ground (highways and rail, including aid to states)	$45,209		Allows state to keep roads, including interstates, in good repair	This is where most pork-barrel spending happens. And roads really should be a state responsibility.	Drivers. Plus, this is a prime congressional "earmark" area. Voters love a new road.
			Trains are environmentally friendly and an important option in the Northeast.	They're not so important in the rest of the country. And Amtrak should pay for itself like any other business.	Business and professional people in the Northeast Corridor— you'll find they're surprisingly loud.
Air, including the FAA, airports, air traffic control, research	$18,005		The country really can't run without reliable air travel.	A lot of this goes to small airports that don't get much traffic.	Anyone who flies. Plus the airlines, and any community with an airport, large or small.
Water, including marine safety	$6,688		Safe shipping is critical to the economy, not to mention recreational boaters.	Shipping lines and boaters could shoulder more of the cost.	Boaters, merchant seamen

PROGRAM In millions of dollars Minus indicates revenue	2006 BUDGET	YOUR BUDGET	PROS	CONS	WHO CARES
Veterans benefits	**2006 Total:** $69,842		Those who served our country deserve to be protected and rewarded	You rarely hear any actual opposition to this—it's mostly a question of whether the system can be more efficient	Veterans, plus anyone who feels strongly about veterans. Which adds up to a lot of people
Income security for veterans	$35,771				
Education, training, rehabilitation	$2,638				
Hospital/ medical care	$29,888				
Housing	−$1,242				
Other benefits and services	$2,787				
Medicare	**2006 Total:** $329,868	$329,868	Okay, if you've been paying any attention at all during the course of this book, you know that you can't just draw a line through these items. But we're leaving them in to remind you what a large share of the budget they represent.		
Social Security	**2006 Total:** $548,549	$548,549			
Income Security	**2006 Total:** $35,247		This is a grab bag of cash benefits to individuals, including federal retirees, the disabled, the unemployed, food stamp recipients, free school lunch, low-income housing aid and others. Essentially you're talking about aid to people who for whatever reason can't completely fend for themselves. But obviously people get this aid for very different reasons—a federal retiree and a child getting free school lunches may have very little in common, and the programs work in completely different ways.		
General retirement and disabitlity	$4,592				The disabled

PROGRAM In millions of dollars Minus indicates revenue	2006 BUDGET	YOUR BUDGET	PROS	CONS	WHO CARES
Federal employee retirement and disability	$98,296				Retirees, current workers, and their unions
Unemployment and compensation	$33,814				People out of work
Housing assistance	$38,295				The poor, and those who care about them
Food and nutrition assistance	$53,928				See above, and add educators for the school lunch program.
Other income security	$123,552				
Energy	**2006 Total: $782**		This includes civilian energy projects like the Strategic Petroleum Reserve, energy conservation grants, the Yucca Mountain nuclear waste facility, and so on. A lot of the costs are offset by fees and profits from agencies like the Tennessee Valley Authority.		
Energy supply	$231				
Energy conservation	$747				
Emergency energy preparedness	–$441				
Energy information policy, regulation	$245				

YOUR TOTAL EXPENSES =

REVENUE SOURCE	2006 BUDGET	YOUR BUDGET	PROS	CONS	WHO CARES
Individual income taxes	**2006 Total:** $1,043,908		It's a tax that falls heaviest on whether people—the more you make, the more you pay.	Even so, nearly everyone pays this and any changes (up or down) resonate through the whole economy.	Everybody
Corporation income taxes	**2006 Total:** $353,915		They've got lots of money.	Raise these too high and companies may cut jobs or pass the cost on to consumers.	You know who
Excise taxes	**2006 Total:** $73,961				
Tobacco	$7,710		Besides bringing in money, this will cut smoking	Anybody who's likely to quit smoking because of the cost has already done it. We're down to the hard core now	Smokers, naturally, and tobacco companies
Alcohol	$8,484		See above, just replace "smoking" with "drinking"	How much of a burden should one industry take on?	Drinkers and brewers— remember, there's a lot more of them than smokers.
Gasoline	–$2,386		Would encourage conservation, fuel-efficient cars, and energy independence.	For many Americans, driving is a necessity, not a luxury—there's only so much cutting back they can do.	Drivers, the oil industry, the auto industry— that covers just about everyone.
Telephone	$4,897				

REVENUE SOURCE	2006 BUDGET	YOUR BUDGET	PROS	CONS	WHO CARES
Other receipts	2006 Total: $97,649				
Estate and gift taxes	$27,877		Besides the money, eliminating this tax means the wealthy families would get their billions tax-free.	This can be a particular burden to families that own small businesses or farms.	People with estates of $1 million or more and their families
Custom duties/fees	$24,810		One of the government's oldest sources of revenue— even predates the income tax		
Federal Reserve earnings	$29,945				
Other misc. receipts (passports, national park entry fees, etc. You could always add more, like a charge for using Global Positioning System).	$15,017		If you need these services, you should pay for them.	Doesn't bring in that much money and might price some people out (do you want to make it harder for poor people to visit a national park?)	Anybody who needs something specific from the government

YOUR TOTAL REVENUE =

YOUR TOTAL EXPENSES =

DID YOU BALANCE THE BUDGET?

IF YOU RAISED TAXES ON "X," HOW MUCH WOLD YOU GET?

What happens on the revenue side depends greatly on how you go about it. So it's a vast oversimplification to just write new numbers in here. Still it's a useful exercise. To get you started, here are some rule-of-thumb estimates you can use for how much money could be raised, courtesy of the Congressional Budget Office.

TAX INCREASE	RAISES IN ONE YEAR (2008 ESTIMATE)
Raise individual tax rates for everyone by 1 percentage point:	$21.1 billion
Raise just the top tax rate by 1 percentage point	$3.9 billion
Eliminate the tax deduction for state/ local taxes	$10.5 billion
Eliminate the child tax credit	$9.2 billion
Increase the cigarette tax by 50 cents a pack	$4.3 billion
Increase alcohol tax to $16 per proof gallon	$4.7 billion
Increase motor fuel tax by 50 cents per gallon	$49.3 billion
Raise fees to cover all food safety inspections	$275 million
TAX CUT	**CUTS, REVENUE IN ONE YEAR (2008 ESTIMATE)**
Eliminating the alternative minimum tax	$34.1 billion
Extending the Bush individual tax cuts	$500 million (or $1.2 trillion by 2017)

Did you still end up with a deficit? Don't feel bad. There's a reason why most deficit-reduction plans are set to work over several years. It's pretty painful to fix the problem all at once. Either the cuts have to be deep or the tax increase steep to make that work.

CHAPTER 17

The "Where Does the Money Go" Voter Protection Kit

Half of the American people never read a newspaper. Half never vote for President. One hopes it is the same half.

—*Gore Vidal*

Unless you've spent the last couple of years on a tropical island talking to a soccer ball, you know that 2008 is an election year, and there'll be another in 2010. And while we don't expect you to cast your vote solely based on the issues we've raised in this book, we hope you'll have them on your checklist as you consider your options. Still, given the nature of politics and campaigns today, that's sometimes easier said than done. People who are running for office quite naturally try to say things that will please as many voters as possible. They also try to avoid saying things that their opponent can use in attack ads or that wind up on YouTube, causing them no end of grief. That's why most of them really don't go into too much detail about how they would address the problem of balancing the budget and making sure that the nation doesn't find itself in financial peril when the boomers start leaving the work force in big numbers.

But the truth is that the people who are elected the next time around—as president, senators, or members of the House—will face hundreds of decisions that will make the country's chances of getting through this in good shape either better or worse. Most of the candidates understand this (and some may even be worrying about it), even though they aren't talking about it much at all.

In this chapter, we've pulled together a few tools to help you feel your way through the political finessing, so you can try to determine whether your candidate will take this problem seriously or not, and if so, what direction he or she will head in to address it. Here's what we have for you.

We start off with our "simultaneous translations" section—what we've labeled as the handy-dandy pocket guide to the political soft sell. The idea here is to help you understand what generally lies behind some oft-heard political rhetoric.

Second, we suggest some steps you can take personally to generate some public energy on this issue. You can act on some of these suggestions in upcoming elections. You can act on the others any old time. Remember, even if the election turns out your way, it's important to keep the pressure on and give your guy or gal the backing he or she will need to get the country moving on this.

Third, we have taken this opportunity to nag just a bit (sorry, we couldn't resist). We've included five signs that you're being a lazy citizen (we've had some moments of laziness ourselves from time to time) and five signs that you're really part of the problems (not you, of course, but maybe someone you know).

Finally, we strongly recommend that you check out the Concord Coalition's list of questions to ask candidates available at www.concordcoalition.org. Even if your YouTube video isn't chosen for a TV debate, you still have plenty of chances to ask candidates about this issue. You can attend one of their meetings in your community, send questions to their Web site, drop by their campaign office and talk to the staff, or drop a note to local reporters suggesting that they check out how the candidates stand on these questions. There are many, many ways to do this.

So here they are:

SIMULTANEOUS TRANSLATIONS—A HANDY-DANDY POCKET GUIDE TO THE POLITICAL SOFT SELL

Politicians are people pleasers par excellence. You can't get very far in politics without learning to talk about policies so most people say, "Umm, that sounds pretty good." Selling ideas and getting people to support them (and you, if you're the politician) is the very heart of politics. There's nothing deeply wrong about it as long as it's not deliberately deceitful or misleading.

Putting your best foot forward is essential in

Spending Uncle Sam's Money (1899)

Concerns about excessive government spending are hardly new. *Credit:* Spending Uncle Sam's Money *by T. Dart Walker (1869–1914), U.S. Senate Art and History Collection*

business; it's crucial in romance, and it's utterly indispensable in political life. In recent elections, the news media and the pundits have often gotten themselves into snickering fits over candidates who can't come up with a "vision that resonates with America today" or who repeatedly "get off message." Voters don't generally reward these candidates, either, so getting the words right is often the difference between political success and failure.

Getting the words right is also often the result of a fair bit of work and not a little amount of money. Politicians gearing up for elections typically hire consultants and conduct focus groups to test different ways of saying things. Once candidates have refined their "message," they and their "surrogates" (their supporters and staff) often speak from "talking points"—those phrases you hear over and over again—campaign stop after campaign stop, interview after interview, speech after speech, until everyone is "sick of it."

Using talking points is common in campaigns and in between elections as well. Sometimes the news media have a little fun with this by showing clips of people all over Capitol Hill using the very same words to describe some hot potato political issue. The perpetrators are either "scripted," or "good at staying on message" depending on your point of view. Personally, we think a little more spontaneity and a little less packaging would be nice, but we'll leave that fight for another day.

SUCH A NICE SLOGAN . . .

The problem for the issue we're talking about right now—the country's looming budget mess—is when voters respond to nice-sounding words and don't bother checking up on what these nice-sounding words actually mean. As we've cautioned before, and we'll caution again, you simply have to read past the headlines. You can't give someone your vote or back their plan just because you like the slogan. Politicians may sell their policies like they were products, but they're not. If you don't like what Congress and the president have done with the federal budget, you can't take it back to the store and exchange it. You have to live with it until the next election comes around. So before you buy into anything, you need to find out what's really being proposed and take the time to decide whether you think it's fair, workable, practical, ethical, reasonable—whether it's something you can live with.

Here's some of the prevailing political soft sell from both Democrats and Republicans, the left and the right. All these ideas and phrases sound utterly wonderful, but if you're smart, you'll approach them with caution. Make sure that you really buy into what they represent. These will seem pretty old hat to you if you're a bit of a policy wonk yourself, but, frankly, we're surprised at the number of Americans who take these lovely little phrases at face value.

"We've got to lick this budget problem by eliminating waste, fraud, and abuse in government." You would have to be a crazy person (or a beneficiary perhaps) not to want waste, fraud, and abuse hacked out of government. There's plenty of it in there. The problem here is that all too often, this phrase allows politicians to pretend that we don't have to do anything else to solve the country's money mess—nothing that will *upset* anyone. We won't have to cut any popular programs. We won't have to raise any taxes. Just clean up that nasty old waste, fraud, and abuse. If you've read every page of this book up to now, you know it's just not that simple. If you've been skimming (it's OK; we've been known to skim ourselves), go back to chapter 10 to get a more realistic view.

"A growing economy—that's the way to fix the budget problem." It's hard to find anyone who doesn't want the economy to grow. Try to imagine anyone arguing that it should shrink. But in today's political life, politicians and elected officials who repeatedly call for "growth" are generally recommending keeping taxes very low and cutting government spending way back. If that's what you back yourself, these are your folk. But if you think protecting some kinds of government spending is important and raising taxes is a reasonable way to pay for the things that matter to you, these may not be your folk. And if you decide on the growth approach, you really need to think about how much growth is possible on a regular basis and whether it will really deliver enough money to cover the budget crunch to come. Remember, the expenses coming up with boomer retirements are huge.

"It's time to invest in the middle class. That's the first priority." Only about one American out of ten considers himself either lower or upper class, so recommending proposals to benefit the middle class is bound to be popular.[1] At the moment, phrases like this are often an indicator that the speaker wants to expand health care coverage and put more government money into education and child care—also very popular. And given that some 45 million Americans don't have health insur-

A Falling Off of Bosses (1881)

And concerns about influence and corruption are not new either. *Credit: Unidentified, after Thomas Nast* Harper's Weekly *wood engraving*

---

[1] According to a May 2006 USA Today/Gallup survey, 42 percent of Americans say they are middle class, 31 percent working class, 19 percent upper middle class, 6 percent lower class, and 1 percent don't know.

ance, a lot of Americans do want to see this gap addressed. The dilemma is that covering uninsured Americans and upping spending on education and child care will obviously cost money; the U.S. government is already in debt, and we have even bigger expenses coming up with the boomer retirements. To our way of thinking, politicians who propose these kinds of "middle-class investments" are honor bound to talk about how they will pay for them and what they plan to do to address the country's longer-term budget problems. And if you're a voter who basically supports this kind of government spending, you're honor bound to think about how to cover the costs as well.

"We want America to be an ownership society." Owning a home is often considered a major step forward in becoming a solid part of the middle class. Nearly all of us can remember the first car we ever owned.[2] Whether we're on the upper or lower end of the income spectrum, most of us take some sense of pride and security in what we own, so it's not surprising that having the country become more of an "ownership society" has a nice ring to it. Politically speaking, this phrase generally indicates that the speaker supports private accounts in Social Security and perhaps some kinds of private savings accounts as a way to reform Medicare. These are legitimate and interesting ideas if they are done responsibly and gradually, with an eye to how they will affect the budget. We cover them briefly in chapter 9. We also encourage you to find out more out how ideas like this would work at the Web sites of groups like Heritage and Cato (which support private accounts) and the American Association of Retired Persons, or AARP (which opposes them). The main point here is to think about how this affects the overall federal budget and the economy. Depending on how they're constructed, private

---

[2] A 1975 Plymouth Duster and a beautiful green VW Beetle, in case you were wondering.

accounts could be a real budget buster (we talk about this in chapter 9). Even if you're convinced you can make out better by investing your own Social Security account, you also have a responsibility to consider how the plan would affect the big picture.

"We have to keep our promises to America's seniors." We're all taught not to break promises, especially to people we love and value, and it would be hard to find anyone— left, right, or center—who really wants to leave a lot of older Americans in poverty and without adequate health care when they are too old and frail to work. The problem with that nice little nice phrase, "keep our promises," is that it sometimes means that the speaker considers any change whatsoever to Social Security or Medicare as verboten. There's an important and genuine difference between treating these programs (or any other government program or any tax cut for that matter) as sacrosanct, and developing policies that honor the society's promise to ensure that old age is not a time of fear and want. So when you listen to a politician talk about "keeping our promises," find out what he or she really means. And if he considers Social Security and Medicare untouchable, it's time to ask him how he plans to pay the bill.

Taking Matters into Your Own Hands

By now, we hope you're concerned enough to want to do something personally about the country's budget problems. A good Wisconsin man we once knew used to talk about being "mad enough to make a rabbit fight a bear." Maybe that's how you're feeling about now. And on this issue, a few hopping mad rabbits might not be a bad thing.

You could run for office yourself of course, and if you

think you've got the determination and tough skin it takes to become a political candidate, here's to you. John Adams and Thomas Jefferson didn't envision Congress as a body of full-time, lifelong politicians with their own political action committees and permanent campaign advisers, so maybe it is time to think about bringing some "less-professional" types to Washington. You could be one of them.

MAKE YOUR CHECK OUT TO . . .

You can also give money out of your own pocket to pay down the debt. The Bureau of the Public Debt, which is part of the Treasury Department, accepts donations to pay down the debt at its Web site. This used to be a stand-alone site that was easy to find and navigate once you were there. But it has recently been moved to the Treasury Department Web site and buried under a lot of information that seems more directed to people buying and selling treasury bonds than to the general public. None the less, it is worth hunting and pecking your way there. Go to www.treasurydirect.gov and follow the links to the public debt section. Want to give? Click on "Gifts." In fact, the bureau received close to $1.5 million in contributions to pay down the debt in 2006. As they so deftly put it, "gifts to reduce debt held by the public may be inter vivos gifts or testamentary bequests." We guess that means they'll take the money whether you're dead or alive.

Whether you're tempted to donate or not, the Bureau of the Public Debt Web site is worth a visit. It has daily "to-the-penny" reports on how much money the country owes, and consequently it's one of the few places you'll ever see the nation's trillions of dollars in debt completely written out. You can also find out how much the country owed every year since 1791. The lowest point seems to be $33,733 in 1835 (that was under President Andrew Jackson, just in case that slipped your mind). The highest, of course, is right now.

But if you're not planning on running for office and can't

quite bring yourself to put money into the government's till given its track record on using your hard-earned dollars to date, there are still important things you can do—you personally.

Get thee to those "meet the candidate" sessions. Ask questions about the budget and the debt and the aging of the boomers. Yes, we said this before, but it is so, so important. We have a presidential election in 2008. A third of the Senate and the entire House of Representatives are up for reelection every two years. In nearly every community, there are campaign events and campaign offices where you can go and talk to people about this budget issue, show your concern, ask questions, and (we hope) give this issue a push up on the political agenda. It's not necessarily easy to talk one-on-one to the presidential candidates or to senators and their challengers in big states. But you can almost certainly talk to candidates for the House. In nearly every district in the country, the incumbents and the challengers spend a couple of months going around the district meeting and talking with potential voters. *This is your chance!* A lot of the events aren't crowded at all; in fact, most candidates have to hire someone to run around town beforehand to dust up participants. Otherwise their boss will end up sitting there talking to five or six people, which can make a candidate very grumpy. Call the candidates' headquarters and/or visit their Web sites to find out when they'll be at a location near you. Ask them about their position on the budget and the debt and how to address the problems facing Social Security and Medicare because of the retiring boomers. And don't let them brag about how they're going to cut taxes or spend more on kids or whatever their schtick is without asking them (politely of course) what their idea would mean for the deficit and the debt.

And one final thought on this. You may think that we're suggesting that you hound the candidates "on the other side" of whichever side you're on, but we're not. The best thing you can do is to make sure that candidates you support and are likely to vote for know that you care about this issue. Politicians are much more concerned about what their supporters think than what their opponents think. And make sure they know that you'll be checking on how they handle it once they're in office as well.

Talk about this issue to people you know. It's vital to talk to people who will be in Congress and other areas of government since they'll be casting votes and making decisions that either move the country toward solutions or make the situation worse. But don't stop with the politicians. You can help build a movement of people concerned about this issue just by talking to the people you know.

When we were writing this book, we had the chance to observe focus groups of typical Americans talking about the country's budget problems. The focus group moderator would generally start out by asking the dozen or so participants to name the most important issues facing the country, and at first hardly anyone mentioned the country's routine deficit spending or rising debt. But when just one person brought the problem up, it really made an impact. That's because a lot of us know in our heart of hearts that it just can't be a good idea for the government to routinely spend more than it takes in. And most of us can quickly see how the country's budget problems affect nearly everything we care about.

Just talking to people you know about this issue may seem trivial, but don't underestimate how much influence you can have. Talking with family, friends, and neighbors is the way a lot of us learn about issues and sort through our views on them. And if you're a boomer parent with kids who will be (or should be) voting soon, you owe them this. They'll be the ones starting careers, trying to raise families, buying their first home, struggling to make ends meet if and when the really bad economic scenarios come into play. You wouldn't send them out into the world without helping them understand the health and safety risks they need to avoid. Make sure they understand this risk as well.

Send information about the issue to your e-mail list. Come on now, we know you're sending those Jay Leno and David Letterman jokes to your friends, or maybe it's recipes and the latest photos of the kids or the cat. We also know that word of mouth about movies, books, TV shows, restaurants, products, and issues spread around cyberspace by people who know each other can be very powerful. Communications and advertising professionals call this viral marketing. (Being able to come up with a new name for "word of mouth" that looks good on a PowerPoint slide is why they get the big bucks and the corner office.) Now, we're not suggesting that you drive your loved ones crazy with a daily blitz on the state of the country's finances. Nor are we suggesting you spam people. But passing along a good article or op-ed on this issue every now and then seems entirely appropriate given the stakes. We've included a list of Web sites that have good information on this issue on a regular basis (see the appendix). Make a habit of visiting them every once in a while, and when you see something interesting and important, pass it along.

Contact journalists who cover the issue—or who should have covered this issue but didn't. A lot of reporters, editors, and news producers think that most Americans don't care about the budget issue, and that if they cover it, most of the audience will think it's boring. If you've come this far in this book, you know that this issue is anything but boring. It's a threat to our country and our way of life, and we need good journalists to give it the attention it warrants. And you don't need to be shy or feel like this is not your place. The truth is that good journalists who report on serious public policy and political issues often wonder whether they're having an impact, whether people are actually reading or listening. Some journalists we've talked with even say that they appreciate hearing from readers, viewers, or listeners about their stories, provided the feedback is reasonable and polite, of course. No one wants to hear from someone who is swearing, screaming, or frothing at the mouth.

Some journalists even obligingly include e-mail addresses in their columns and shows, so that's your invitation. Contact reporters, editors, producers, and webmasters to suggest more coverage on this issue. Contact them when they cover proposals for new spending or for tax cuts and suggest that the next time they do a story like this, they include information about how the idea would affect the budget and the debt. Just as important—maybe even more so—take a moment to compliment good coverage when you see it. Like anyone else, journalists like to know that someone appreciates the work they do. And then they can show the higher-ups that someone does indeed care about this issue.

Become a maven. In his influential book *The Tipping Point*, Malcolm Gladwell describes the impact that individuals who care a lot about an issue have on society's ability to make progress. "Mavens" don't have to have PhD's in the subject at hand; they don't have to hold elected office or appear regularly on CNN or Fox. They don't have to be wealthy or powerful. They do have to care about an issue and be willing to put in some work to try to promote change.

If you care about what you've been reading, and you're ready to roll up your sleeves, there is quite a bit of help out there. One way to start down the road to mavenhood is to organize a meeting among people you know or with an organization you're familiar with to talk about this issue and trade ideas about how to address it. A number of nonpartisan organizations have materials, moderator training, and other kinds of help for you to do just that. Full disclosure requires us to say that our organization, Public Agenda, has worked with these groups in the past, and we know the key people involved in them. And having worked with them, we can recommend them with confidence.

One is the **National Issues Forums (NIF),** a national coalition of colleges and universities, schools, clubs and service organizations, churches, and other local institutions that sponsor nonpartisan discussions on public issues. As part of their work to get people talking, they publish citizens' guides, each of which describes three or four alternative approaches to the issue it covers, complete with pros and cons. Not too long ago, they have published a guide like this on the deficit and the debt called *The $9 Trillion Debt: Breaking the Habit of Deficit Spending*, by Keith Melville (available at www.nifi.org). So what would a future maven do? Call NIF (1-800-433-7834) or visit its Web site (www.nifi.org) and find out how to organize a citizen discussion in your area yourself or how to work with an organization to do it. NIF has helped scores of people organize meetings around important issues, so here's your chance.

Another option is to think about joining the **Concord Coalition.** It's a bipartisan group based in Washington, D.C., devoted to putting this issue in the public spotlight and keeping the heat on. Concord publishes updates on what's happening on this issue, and their Web site (www.concordcoalition .org) features an online "penny" version of the nation's revenues and spending so you can check on how much you've learned reading this book. Concord also participates in and helps organize the Fiscal Wake-up Tour—a series of town meetings with an extraordinary panel of budget experts. It's a riveting presentation, a minicourse in the budget issue, and hearing people from the Concord Coalition, the Brookings Institution, the Heritage Foundation, and the Committee for Economic Development describe why they are so concerned really does get the juices going. A budding maven would certainly want to attend the Wake-up Tour. A full-fledged maven would talk with Concord about how to get the Wake-up Tour to come to his or her city.

Plus, there's another good option even if we do say so ourselves. We think it's time for you to become a regular contributor and accomplice at **Facing Up to the Nation's Finances** (www.facingup.org). Facing Up is a joint enterprise of Public Agenda (our organization), Concord (see above), Viewpoint Learning (a research organization), and the Heritage and Brookings think tanks. We're hoping it will become the Grand Central Station for all things budgetary. And unlike a lot of Web sites covering the country's budget debates and challenges, this one is specifically designed for the public. It will give you choices, not opinions, and explanations, not expert minutiae. It's a place where you can discuss budget issues, sign up for regular reports, download discussion guides, and even work on solutions with other people who care about this problem. We plan to be doing a bit of blogging there ourselves, so here's your invitation to join us.

FIVE SIGNS YOU'RE BEING A LAZY CITIZEN

In *Where Does the Money Go?* we've been fairly tough on the nation's leaders. There are some fine and honorable individuals in government these days (we'll even go out on a limb and say that there are actually a fair number of them), but as a group, our leaders have let us down.

But we, meaning the public at large, haven't asked very much of them, have we? There's no reason to feel either smug or victimized. Not enough of us pay attention to politics, much less vote or get active. A lot of us just are not doing as good a job as we need to do when it comes to being responsible citizens.

A lot of this is about attitude. In our view, too many people these days can't tell the difference between good government and good customer service. You can get good customer service at the drive-through window at McDonald's: you pull up, talk to the squawk box, and pull out with your supersized helpings of high-fructose corn-syrup-related products.

Government isn't that simple. Oh sure, it's fair to expect the government to process your passport renewal or building permit application promptly, maybe even as promptly as McDonald's gets out hamburgers. The difference is that we're not expected to take an active role in how McDonald's is managed. The company tries to respond to what customers want, but it's not holding elections every couple years to decide whether the bacon stays on the Egg McMuffin. The federal government, in its way, is waiting for that kind of signal.

We know, we know. Life is hectic these days; there's not enough time to do everything we should be doing to earn those good "doobee" awards. But some of us have gotten a little too lazy, a little too complacent in the citizenship department. Could that be you? We've listed five signs to look out for, and if any of these sound a little familiar, we hope you'll think about putting in some extra effort. We really, really need you back.

1. You're not registered to vote (or you've let your registration lapse). Politics is not a spectator sport. You can't just watch the political debate from the sidelines and boo when things don't go your way. And yes, we know, sometimes the choices among the candidates can be pretty uninspiring. But this low-voter-turnout thing has allowed very small groups (special interests, lobbyists, big political donors) to have far too much say in what the country does. We've got to get in there and fight the good fight. So if you're not registered, turn on your computer and go to www.vote411.org this very moment. Follow the directions there, and get it done. And then, please, please inform yourself about the candidates and cast your vote. In the lottery of life, being born in the United States is a big, big win. Nearly every person in this country is safer, warmer, and better fed and has more opportunities than billions of people around the globe. There's just no excuse for not fulfilling this small obligation of citizenship.

And, by the way, that includes not being deterred by long lines, broken voting machines, or other Election Day hassles. Millions of people are willing to grouse bitterly as they keep hitting "redial" to cast a vote for *American Idol* and yet won't sacrifice a few more minutes standing in line to pick the next president. The League of Women Voters' VOTE411 site gives you the basic information you need to make sure you're not turned away from the polls. Once you decide who you're going to vote for, don't let anything keep you out of that booth.

2. You vote only in presidential elections. The presidential election is the big kahuna, no doubt about that. But on this budget issue, no president can fix the problem by him- or herself. And if we happen to get a president next time round who's not paying attention to this issue, you're going to want people in Congress to stop him or her from doing even more financial damage. So your senators and representatives are essential players in resolving this mess. In most states, there are multiple opportunities to go somewhere and actually speak to these people directly. Do it. Find out about them and their stands. Talk to them about the budget. We're not sure whether it's a good thing or a bad thing, but most senators and representatives are essentially running for office all the time. Take advantage of this. Give them a piece of your mind (politely, of course), and give them your support at the ballot box when they deserve it.

3. If they agree with you on X, they've got your vote. Many Americans have an issue they really care about. Abortion and the Iraq War are two that come to mind. Naturally, we all look for candidates who agree with us in our "most-important" area. That's entirely legitimate, and we would never suggest that you support candidates who represent the "wrong" view in an area you're genuinely passionate about. But in most cases—and especially in the primaries—there are several candidates whose views generally fall on your side of the issue. You really need to investigate how these people look at the other important issues the country faces. Just because they're good on X, that doesn't mean they're good on everything else. And—we're just asking, of course—you might want to support some candidates who are good on your issue, even if they're not especially eloquent or dramatic about it, because they have some important things to say in other areas. The country has a lot of very important decisions coming up in the next several years. We need people in political office who can operate effectively on a number of fronts.

4. You never listen to candidates' debates. Sure, you can get some idea of what candidates and elected officials stand for by watching or reading the news. And if you want to find out how truly wonderful they all are—every last one of them—you can watch their ads, go to their Web sites, and read their campaign literature. But if you want to find out what they think and why, and how it differs from what the other candidates think and why, your best bet is one of the debates. There are more and more debates taking place, and based on what we've seen recently, the journalists who moderate them and try to move things along (it's usually reporters, anchors, and the like) are doing a better job of focusing on important issues. So while you don't have to watch all the debates in their entirety (that probably is for political junkies), you should watch at least one. And don't just look for the gaffes (most politicians do say something relatively stupid at one point or another, as do most of the rest of us). Use this opportunity to compare and contrast. Think about who has the best ideas and is most likely to follow through on them.

5. You just read the headlines. OK, at least you're glancing at the news, which is definitely commendable. But particularly on this budget issue, that's not enough. You really need to read at least a few paragraphs to find out what's really happening. The headline screams "Big Tax Hikes Proposed," and it turns out to be a proposal so modest that you won't even notice that your taxes have changed come April 15. The headline screams "Candidate Backs Social Security Cuts," and it turns out that he or she is really talking about changes that affect only high-income recipients. You can make up your own mind about whether the proposal is really something you can live with, but do it based on the facts—not on a headline someone wrote just to get your attention.

FIVE SIGNS YOU'RE PART OF THE PROBLEM

When it comes to being good citizens, there's another group of Americans who are not just lazy—they've checked out entirely. The news? It bores them. Politics? They can't be bothered. Elections? "Come on, man. I'm just not into that stuff." Most of us know someone who falls into this category—those people are not exactly few and far between. We would be amazed if any of those so-called citizens actually read this book, but if you're feeling a little out of sorts about people who take the country for granted while they enjoy its benefits, you could show them this list. Maybe it will at least induce a smidgeon of guilt.

1. You can name every party to the legal entanglements surrounding the death of Anna Nicole Smith, but you don't recognize the name of the vice president or the

Speaker of the House. We're prepared to let you slide on the names of the secretary of agriculture and the surgeon general. We're even willing to give you a pass if the names of the vice president and the Speaker are on the tip of your tongue, but you just can't quite come up with them right now. But the fact that the E! network does not hire Joan and Melissa Rivers to do red-carpet fashion snark at the State of the Union address is no excuse for not knowing who the nation's leaders are.

2. You're getting all your news from comedians. How can you even understand the jokes without knowing what actually happened? This seems strange to us, but apparently late-night TV is as close to the news as some people get these days. Sorry, folks, it's just not enough. And as much as we love Jon Stewart, and as intelligent and thought-provoking as his interviews are, you still need to keep up with the *real* news. It's so easy now, with all the online sources available. Really, there's just no excuse.

3. You watch every single game in the NCAA tournament, but you don't have time to keep up with politics. OK, do you sense a theme here? We're talking about people who'd rather be amused than informed. There you are watching sixty-five teams fight it out during March Madness (and who knows how many baseball games in the summer and football games in the fall), and you say you don't have time to watch the news or participate in an election. Sorry, this just doesn't cut it with us. We like sports. Sports are a fine thing. But if you can make time to be a fan, you can make time to keep up with what's happening in the world and in our country. You can find the time to fulfill the most minimal role of a citizen.

4. You're just focused on your family. Politics doesn't matter to you. Most people who say things like this have one or more lovable little creatures at home, and they absolutely should be the focus of your attention the lion's share of the time. But if you really care about their future, you better start paying attention to politics as well. Right now, this very year, elected officials are making decisions that will determine what kind of economy and government the next generation will have. You can make kids zip up their jackets and wear bike helmets, but they're still heading off into a world where crushing financial obligations are going to make their lives much, much harder than necessary. If you stick with that "politics doesn't matter line," you're really not doing those little kids any favors.

5. The news is depressing. You'd rather not know. All right, you've got us there. The news often is depressing. It can also be frustrating, even infuriating. Sometimes it's even scary. But just because you're not paying attention doesn't mean it's not happening. In this day and age, the ignorance-is-bliss approach is selfish, stupid, and dangerous. If enough of us run off to some "I'd rather not know" la-la land, it could eventually bring us all down.

★★★★★★★★★★★★★★★★★★★★★★★★★★

CHAPTER 18

The Last Word: Six Realities We Need to Accept to Solve This Problem

I want you to get up right now, sit up, go to your windows, open them and stick your head out and yell—"I'm as mad as hell and I'm not going to take this anymore!" Things have got to change. But first, you've gotta get mad!

—*Howard Beale (Peter Finch) in Paddy Chayefsky's screenplay of* Network, *1976*

A couple of hundred pages ago, we started off with "The Six Points You Need to Know to Understand the Federal Budget Debate." Our goal was to explain why this issue is so important and help you start thinking about the direction you want the country to take to solve the problem.

Along the way, we expect we've occasionally aggravated readers who are genuine budget experts by "oversimplifying" the issue. Yes, we admit it—this is a once-over-lightly treatment. At the same time, we've probably perplexed quite a few readers who aren't budget pros at one point or another. We were occasionally baffled ourselves while we were writing this, so there's no reason to expect that other people won't be as well. Some of the details are extremely complex and confusing. Still and all, if you just stop multitasking for an hour or so, if you just concentrate on what we've laid out here, we're pretty sure you can get a handle on this—enough of a handle to help yourself and your country out.

The U.S. Senate in Session (1874)

The United States has a representative government, but it is our responsibility to vote and vote wisely. *Credit:* The United States Senate in Session *by unknown artist,* Harper's Weekly *wood engraving, US Senate, Art and History Collection, and 108th Congress by U.S. Senate Photo Studio*

If you haven't gotten all the numbers down pat yet, don't worry. The numbers change all the time anyway—it's the big picture you need to keep an eye on. If you haven't pinned down your own list of best tax and spending policies yet, don't worry about that, either. Frankly, we're not so sure that any of us should have a list like that right now—especially not one that's carved in stone like the Ten Commandments. Chris Edwards over at the Cato Institute has rather bravely published a list of very precise cuts that would whittle the budget down to size (look for *Downsizing the Federal Government,* available at www.catostore.org). You may not like all of them (if you're typical, you probably won't like all of them), but we give the Catos credit for putting some specifics on the table. The people over at Brookings have suggested alternative ways—roughly left, right, and center—for tackling this problem. It's part of their Fiscal Sanity project. There's plenty for you to think about there (see the Fiscal Sanity books for 2004, 2005, and 2007, available at www.brookings.edu/press/bookstore.htm). And the gurus over at Heritage offer a virtual budget library on their Web site at www.heritage.org/research/budget. Flip through Stuart Butler's or Alison Fraser's presentations for the Fiscal Wake-up Tour, and you'll definitely be able to wow your friends with some eye-opening budget projections. We've added some tools of our own—our own little budget work- sheet (chapter 16) and a "greatest hits" list of alternative ways to tackle the Social Security and Medicare issue (chapter 9).

JUST DON'T GET TOO ATTACHED

Be forewarned though. Even if you carefully work your way through every single one of these tools (and others), you may still feel uncertain about exactly what the coun- try should do. And that's just fine from our point of view. It's even preferable as far as we're concerned. If we've all made up our minds—if we've all got our lists of deal break- ers—the country will end up spending the next couple of decades fighting it out. If enough of us remain open-minded and flexible, if enough of us are still prepared to listen and adapt, the country will be in a better position to find the compromises to work our way out of this fix.

So from our point of view, a little indecision is an entirely good thing, but there are some things that really shouldn't be up in the air. We close the book with another set of six ideas—in this case "Six Realities We Need to Accept to Solve This Problem."

Reality No. 1: We have to start now. There really is no time to lose. Every year we have a budget deficit just bulks up the debt and limits our options. Postponing discussions about what to do about the problems facing Social Security and Medicare won't make them go away. No matter which kinds of solutions we choose, they will be easier financially and politically if we start now.

Reality No. 2: We have a short-term problem and a long-term problem—we need to address them both. Nearly every year for nearly four decades, the U.S. government has spent more money than it takes in, and finding a way to balance the budget going forward is the central short-term issue. But it's not enough. We also have to face up to the longer-term problem of how to pay for Social Security and Medicare for the aging boomers. To our way of thinking, we simply have to get going on both of these problems—they're interrelated and intertwined. There are a few talking heads out there who say that balancing the budget every year is not important, that we just need to tackle the Social Security and Medicare issues. But given the size of the debt and the money the country is paying in interest payments, combined with the humongous bills we have coming due very soon now, we just don't buy it.

Reality No. 3: We need to address the waste, fraud, and abuse issue, and then we need to move on. While we were writing this book, we were often angry at the way people in Washington misuse the public's money. Check out the news almost any day of the week, and you'll find examples of government waste and inefficiency, of decisions being made for the few and not for Americans at large. Lavish spending at the Smithsonian; unconscionable waste on reconstruction projects in Iraq; bills written by lobbyists—we could go on, but others will do it for us. It's more than time to go after the waste and carelessness that characterizes way too much of what Washington does. But once we put reforms into place, we have to move on. We have to let this problem go. No organization, no business, no household ever operates with pure efficiency—we do have human beings involved here, after all. What's more, even slashing everything that could conceivably be considered wasteful or problematic just won't solve the overarching problem. We have some tough decisions to make. We need to face up to them.

Reality No. 4. We need voters to demand that candidates take a stand on this issue. It's a cop-out to get mad about what Congress and the administration do if we're not willing to hold them accountable. We need to stop voting for candidates who promise new programs or tax cuts without specifying how they will pay for them. We have to stop letting candidates slide by with vague answers about "balancing the budget and cutting the deficit in X years." That's just not enough. We need to demand that candidates and elected officials start talking frankly and realistically about Social Security and Medicare. If we don't do our part as citizens and voters, we will leave the next generation with a horrendous mess not of their making.

Reality No. 5: We need to think about what we can live with—not what we personally want. Every one of us has ideas about what government should spend more on, what it should cut, who should pay more taxes, and who should pay less. Chances are we're just not going to get our way. We simply have to be prepared to compromise and make adjustments in our plans and our thinking. We can and should have a good full-out national discussion over the major pieces—tax policy, how to balance the budget, how to change Social Security and Medicare. Let's have some good old-fashioned arguments about a whole range of ideas. But we can't take our ball and go home if the political winds don't blow our way. Not doing anything about this problem is by far the worst solution of all.

Reality No. 6: To solve this problem, we need a different state of mind. Diet experts often point out that losing weight (and keeping it off) requires a whole new way of thinking about food and a whole new set of eating and exercise habits. It requires a new state of mind. And that's what we need to solve the country's budget problems. We simply have to accept that the country has been overspending and over-promising for years and that it's just not right to keep on shifting our expenses onto the next generation. When the big controversies erupt—what to do about the Bush tax cuts, how to reform Social Security—we need to face them with this mind-set. When we vote, we need to look for candidates who share that mind-set, who will bring it to every decision they make. We're not going to solve this problem in one big debate where we all gnash our teeth and then it's over. Solving this problem depends on thousands of decisions big and small that we will make over the next couple of decades. Unless we face these decisions with a different state of mind from the one we've had in the past, we'll just slide back into our foolish, free-spending ways.

We're convinced the budget problem can be solved. Americans have faced challenges like this before, and we were able to address them. After World War II, for example, the country's debt was higher than the country's entire gross domestic product, and yet the country recovered and eventually prospered.

Today's problems are different, but the country is wealthy and productive. The answers are not easy, but they are not horrifying, either. No one is talking about throwing older people out of hospitals when they are ill or leaving younger ones to fend for themselves when their time comes to retire. Many elected officials are willing to take this on if we support them. And we sense that much of the American public is ready to think about this problem seriously as well. But addressing this problem—making better decisions in the thousands of questions and turning points that will come up—means creating a different political discourse from the one that prevails now.

If there's one hope we have for this book, it's that our readers will begin to look at every single discussion about spending and taxes, about Social Security and Medicare, with a different and more critical eye. We all need a little refrain in our heads: What can we afford? Where will we get the money? What does this mean for the future? Are we being fair to those who come after us? We believe that if the American people start asking themselves these questions, the country will soon be on a more responsible path.

APPENDIX

Places to Go, People to Meet

1. THE SWEET SPOTS FOR INFO

Here are some of the Web's best places to find out more.

Office of Management and Budget (www.whitehouse.gov/omb/
budget)
This is where you will find the president's annual budget
request and rationale for spending the country's money in
the coming year. Not surprisingly, the OMB puts the best foot
forward on the president's plans, but this is the place to go if
you want the details of both the current budget and what's in
the works. Even though it's part of the government, the Office
of Management and Budget has several Web resources that
ought to be useful for budding reformers.

As of this writing, the **OMB Earmark Database** is
the best way of trying to track down earmarks—the little
pork-barrel projects quietly tucked away in the budget.
Unfortunately, the database has some serious limitations.
For example, the site can't tell you who sponsored the earmark (very critical) or even the ultimate beneficiary of the

funding (even more critical). And, by the OMB's definition, earmarks are something only Congress does (although the White House has plenty of ways of serving pork, too). But you can track by agency and state where the money is supposed to be spent. The site is at http://earmarks.omb.gov.

ExpectMore.gov is the OMB Web site devoted to performance ratings for government programs—which ones work and which need improvement. According to the OMB, the vast majority of government programs are "adequate" or better. But fully one-quarter are rated as "not performing," with 3 percent "ineffective" and another 22 percent rated with "results not demonstrated." That means there isn't enough data to prove things one way or another. That category includes some high-profile agencies, such as the Border Patrol and the Equal Employment Opportunity Commission.

Congressional Budget Office (www.cbo.gov)

The CBO is an independent, nonpartisan agency set up to give Congress reliable budget estimates. It's nonpartisan and highly respected, but we're not going to kid you. A lot of CBO reports are a pretty rough read for—well, probably for anyone who's not a budget analyst. When we were using the site, we noticed that the most accessible material from the CBO is generally found in testimony presented before Congress. You might want to have a glass of water handy, because they are really, really dry. But they cover the more subtle aspects of these issues that you just don't find on the evening news.

Government Accountability Office (www.gao.gov)

Formerly known as the General Accounting Office, this is the federal government's auditor. This agency is about as independent as federal bureaus get—its boss, the comptroller general, is appointed to a fifteen-year term, and the GAO is ferociously nonpartisan. GAO auditors review the operations of every government agency and issue often-stinging reports on how the government could function better. The current comptroller general, David Walker, has taken on the task of raising public awareness of the "fiscal tsunami" awaiting us, so this is also an excellent place to find no-holds-barred assessments of the problems.

Public Debt Online (www.treasurydirect.gov/govt/govt.htm)
The government's Treasury direct site includes a daily report on what the country owes down to the penny, along with historical information about the debt and more specialized information for financial types on buying and selling governments bonds. And just in case you're motivated, here's where you can make your voluntary contribution to help pay off the debt.

Monthly Treasury Statement (www.fms.treas.gov/mts)
This site lists the government's income and expenses for the last month, the last year, and the year before that. It's a little more headache-inducing (you can download the monthly statement in Excel, which should give you a little warning), but it is chock-full of specifics on just where the country's money comes from and where it's spent, and not just in the major categories.

Fedspending.org
Run by OMB Watch, this is a searchable database of federal grants and contracts. You can search by department, contractor, state or congressional district, and whether the contract involved competitive bidding or not. It's a great resource for looking at the generally less-examined question of who the federal government pays to do its work. By the time this book comes out, OMB is expected to launch a

government site, **FedSpending.gov,** which may cover the
same territory.

2. GUIDES TO THE ISSUE—LET THEM EXPLAIN IT ALL TO YOU

The $9 Trillion Debt: Breaking the Habit of Deficit Spending, by
Keith Melville, for the National Issues Forums (NIF), May
2007. Written specifically to help citizens think about alternative ways to solve the country's budget problems, *The $9
Trillion Debt* is one of a series of NIF guides on major issues
(visit www.nifi.org for information about the guides and the
discussion forums where they are used). It discusses three
approaches to the budget issue, one focused on trimming
Social Security and Medicare spending, a second exploring
other spending cuts, and a third looking at the option of
raising taxes. The NIF books on the budget and on other
issues include pro and con arguments for each approach,
and all are reviewed by experts with different points of view
for balance.

Restoring Fiscal Sanity is actually a series of books published by the Brookings Institution addressing the same
issues we take up here. But to get an overview of the budget
challenge, you can't do better than the first in the series,
Restoring Fiscal Sanity: How to Balance the Budget (Alice M.
Rivlin and Isabel V. Sawhill, Brookings Institution, 2004).
Subsequent editions of *Restoring Fiscal Sanity* update the
discussion and take a closer look at the particular problem of health care costs. You can download the entire book
from Brookings at www.brookings.edu/es/research/projects/
budget/fiscalsanity.htm.

Federal Revenue and Spending: A Book of Charts, by Alison
Acosta Fraser, Rea S. Hederman Jr., and Michelle Muccio

(Heritage Foundation, 2006). The Heritage Foundation, one of the best-known conservative think tanks, regularly revises and updates this series of charts describing the major trends in the federal budget. It's a good place to start for an overview of the problem. Available online at www .heritage.org/research/features/BudgetChartBook/.

3. GROUPS WORKING ON THE ISSUE

Committee for a Responsible Federal Budget (www.crfb.org). This is a nonprofit, nonpartisan group that is "committed to educating the public about issues that have significant fiscal policy impact." The group is cochaired by William Frenzel, who headed up President Bush's commission on Social Security, and Leon Panetta, who was budget director and later chief of staff for President Clinton. The Web site features interesting speeches and testimony from committee members, updates on the country's fiscal health, and *The Exercise in Hard Choices* workbook for citizens on how to solve the problem.

Concord Coalition (www.concordcoalition.org). Concord was founded during the early 1990s as an organization "advocating fiscal responsibility while ensuring Social Security, Medicare, and Medicaid are secure for all generations." The Concord folks are "budget hawks"—they want action on this issue—so they do have a point of view, but the group is widely respected for its bipartisanship and strong command of the facts. According to its Web site, Concord's "national field staff and loyal group of volunteers cover the country, holding lectures, interactive exercises, conducting classes, giving media interviews, and briefing elected officials and their staffs." The Concord Coalition is a sponsor of the Fiscal Wake-Up Tour, a series of town-hall meetings about this

problem occurring around the country, in partnership with
the Brookings Institution, the Heritage Foundation, the
Committee for Economic Development, and the Committee
for a Responsible Budget.

Facing Up to the Nation's Finances (www.facingup.org) is a
nonpartisan effort to get people talking about this issue
and to have an honest dialogue about how to solve it.
The partners include Public Agenda (the organization we
work for), the Concord Coalition, Brookings, Heritage,
and Viewpoint Learning—groups that agree on very little
other than that this problem is real and serious. In addition to being a great source of information, Facingup.org
is where people who care about this issue can discuss, get
organized, and find out about Facing Up events in their
community.

4. POINTS OF VIEW

In *Where Does the Money Go?* we haven't taken positions on
the various solutions for addressing the country's financial
problems, but there are plenty of think tanks and advocacy
groups that have. We hope, by now, that you'll find yourself curious about what they have to say. Here are some of
the most important that have done significant work in this
area.

American Association of Retired Persons
With over 35 million members, AARP is generally considered the country's most influential organization representing older Americans. In the "Issues and Elections" section
of the AARP Web site (www.aarp.org), you can read the
organization's position papers on Social Security and Medicare, among other related issues.

American Enterprise Institute

One of the country's leading conservative research organizations, its Web site (www.aei.org) is a gateway to AEI's substantial library of reports and publications, many of which address budget and tax issues and Social Security and Medicare. AEI is home to a number of respected budget experts, such as Kevin Hassett, R. Glenn Hubbard, and Daniel Shaviro, along with Mark B. McClellan, former head of the FDA.

Brookings Institution

Brookings (www.brookings.edu) has long been one of Washington's dominant think tanks, home to a banner list of scholars. Just a small sampling of Brookings publications on this theme includes *Saving Social Security: A Balanced Approach*, by Peter A. Diamond and Peter R. Orszag (Brookings Institution Press, 2005), *Social Security and Medicare: Individual vs. Collective Risk and Responsibility*, edited by Sheila Burke, Eric Kingson, and Uwe Reinhardt (Brookings Institution Press, copublished with the National Academy of Social Insurance, 2000), and the *Restoring Fiscal Sanity* series of publications mentioned on page 312. The 2007 edition of *Restoring Fiscal Sanity*, edited by Alice M. Rivlin and Joseph R. Antos (Brookings Institution Press, 2006), focuses on health care costs.

Cato Institute

Even if you come from the liberal side of the aisle, you have to admire the libertarian Cato Institute for being willing to get specific. Cato's *Project on Social Security Choice* (www.socialsecurity.org/catoplan) and *Downsizing the Federal Government: A Blueprint for Federal Budget Reform* (www.cato.org/pubs) provide some of the most detailed information available about private accounts, Social Security, and potential cuts throughout the federal budget.

Center for American Progress

Headed up by former Clinton chief of staff John Podesta, Center for American Progress says that it is working to create a "long-term, progressive vision for America." If you're wondering about the liberal/progressive stance on ideas like raising the retirement age, repealing the estate tax, the priorities of President Bush's budget, and many other related topics, this is the place to go (www.americanprogress.org). The site also offers a regular budget blog.

Center on Budget and Policy Priorities

The center is one of the few nongovernmental research organizations focusing specifically on budgetary policy at both the state and federal level. It's generally categorized as a liberal organization, but like Cato, the Center on Budget and Policy Priorities doesn't shy away from specifics. The organization provided detailed analyses criticizing President Bush's plans for private accounts in Social Security and has created a "Tax Cuts: Myths and Realities" section on its Web site (www.cbpp.org).

Club for Growth

The Club for Growth (www.clubforgrowth.org) believes "prosperity and opportunity come through economic freedom." The club recommends candidates who support its policy goals—extending the Bush tax cuts, personal accounts in Social Security, repealing the estate tax, and others—and sponsors a Club for Growth PAC to provide campaign funds. The site has fresh commentary on current debates, including articles and op-eds from Lawrence Kudlow, a member of the club's economic policy council.

Heritage Foundation

For more than thirty years, the Heritage Foundation has been an influential and respected conservative source of research to "formulate and promote conservative public policies based on the principles of free enterprise, limited government, individual freedom, traditional American values, and a strong national defense." The Heritage Web site (www.heritage.org) offers scores of research reports, issue backgrounders, and commentaries on the issues introduced here.

OMB Watch

OMB Watch (www.ombwatch.org) describes its goal as increasing "government transparency and accountability; to ensure sound, equitable regulatory and budgetary processes and policies; and to protect and promote active citizen participation in our democracy." OMB Watch contains plenty of information about budget developments on the Hill and the status of various budget-related debates. The perspective is liberal, it's fair to say, and recent publications focus on how regulatory agencies such as the Environmental Protection Agency are functioning in the Bush administration.

5. GOOD READS

The Coming Generational Storm: What You Need to Know about America's Economic Future, by Laurence J. Kotlikoff and Scott Burns (MIT Press, 2005). The authors lay out the policy challenges posed by the sheer size of the baby boom. By 2030, the authors point out, walkers will outnumber strollers and Social Security and Medicare will be treading water or worse. Kotlikoff and Burns discuss a number of policy solutions and include what they term a "life jacket"—advice for individuals to see them through the tough times ahead.

Do Deficits Matter? (University of Chicago Press, 1997) and *Taxes, Spending, and the U.S. Government's March toward Bankruptcy* (Cambridge University Press, 2007), by Daniel Shaviro. A professor of law and taxation at New York University, Shaviro has written extensively about budgetary issues with his main focus on the long-term impact of today's spending patterns. Shaviro's blog, Start Making Sense (danshaviro.blogspot.com), also frequently touches on taxation, budget, and entitlement issues.

Running on Empty: How the Democratic and Republican Parties Are Bankrupting Our Future and What Americans Can Do About It, by Pete G. Peterson (Picador, 2005). A former secretary of commerce, cofounder of the Blackstone Group, and board member of Public Agenda, Peterson is a longtime critic of routine deficits and unrealistic government promises. *Running on Empty* is actually the latest of a series of Peterson books on the topic. In *Running on Empty*, Peterson charges that the country's current leadership has "presided over the biggest, most reckless deterioration of America's finances in history."

"Forgive Us Our Debts," by Andrew L. Yarrow (Yale University Press, 2008). Yarrow (who has recently joined Public Agenda in its D.C. office) covers the history of the debt in the United States, the mechanics of federal financing, and the possible consequences, both for individuals and society at large, of not addressing the problem. The book also describes reforms that would help, in Yarrow's words, "to get us out of the woods."

6. BEFORE YOU VOTE

Project Vote Smart provides biographical information, voting records, interest group ratings, and campaign contact information for candidates for national and state offices. Vote Smart has a good track record for nonpartisanship and providing specific, helpful information for voters. Definitely worth a visit at www.vote-smart.org.

Open Secrets works to give a direct, detailed answer to any voter's request to "show me the money." This is the Webby Award–winning site of the nonprofit, nonpartisan Center for Responsive Politics, a group that focuses on money and politics. Enter the name of your member of Congress, and you'll find out how much money he or she raised in the last election and how much came from business, labor, PACs, and so on. This is definitely the place to go if you want to know who is paying the way for the candidates you're considering. It's at www.opensecrets.org.

Maplight.org tries to go Open Secrets one better (in fact they use Open Secrets' data) by tying campaign contributions to legislation. You can search for bills, find out which lobbying groups and organizations favored or opposed the legislation, then see how the contributions match up with the way legislators voted. They're funded by the Sunlight Foundation, the Wallace Alexander Gerbode Foundation, and the Arkay Foundation.

FactCheck.org specializes in helping voters sort out the truth in campaign ads, campaign speeches, and other election sloganeering. You can sign up for a regular news feed, which might be a good idea during the campaign season when the going gets rough, and the truth begins to suffer as a result. Fact Check operates out of the Annenberg Public Policy Center at the University of Pennsylvania. We consider it a must-visit every campaign season.

We're almost to the end of the book now, so it's OK to admit something: You're not sure who your member of Congress is. Not to worry; many people aren't. But this is easily fixed.

A number of sites offer zip code searches to help you find out who represents your town in Washington, but the official congressional sites, **www.house.gov** and **www.senate.gov,** will do just fine. If you want to find out ways of lobbying your representative—petitions you can sign, addresses you can write to—you can try **Congress.org**, a private site that offers a lot of that information. And to track bills in Congress, search voting records, and get basic information about how the legislative branch works, you can't beat the Library of Congress's Thomas site, named for Thomas Jefferson, who surely would approve of empowering people via the Internet, if he was still around. You can find the site at **Thomas.loc.gov.**

Remember that old line about how 90 percent of success in life consists of showing up? It's certainly true on Election Day (unless you live in vote-by-mail Oregon, in which case 90 percent of success consists of having a postage stamp). But for the rest of us, if you're going to make a difference, you've got to register and then show up at your polling place on Election Day. If you've got basic questions about how to register, where your polling place is, and what to do if someone challenges your right to vote, the **vote411.org** Web site run by the **League of Women Voters** has the answers.

★★★★★★★★★★★★★★★★★★★★★★★★★★★

ACKNOWLEDGMENTS

There are many, many people who advised and aided us while we were working on *Where Does the Money Go?* Ruth Wooden at Public Agenda sparked the idea of us writing a book, and Deborah Wadsworth from our board helped us transform a vague notion into a reality. In this case, it is absolutely true that we would not have completed this book without them. All of our Public Agenda colleagues have been enthusiastic, encouraging, and helpful. We would especially like to thank Claudia Feurey, Michael Remaley, and David White for their work in promoting and fine-tuning the book.

Jenny Choi and Nancy Cunningham were our indispensable fact checkers. We have benefited from the intelligence, professionalism, and good humor of Jud Laghi and Larry Kirshbaum at LJK Management, and Matthew Benjamin, Sarah Brown, and Helen Song at HarperCollins. Working with them has been a wonderful experience. We would also like to thank James Capretta, Keith Melville, Alice Rivlin, and Andrew Yarrow, who read drafts of the book and gave us enormously helpful advice and counsel.

Daniel Yankelovich's resounding belief in the good sense of the American public lies behind every word we have written, and for his insight and inspiration, we thank him.

Our families and friends have been remarkable—tolerant when we were distracted and encouraging when we were tired. We especially want to thank Susan Wolfe Bittle and Josu Gallastegui for their unfailing love, support, and patience—not to mention their willingness to read yet another draft without rolling their eyes.